INTERNATIONAL
ENERGY AGENCY

WITHDRAWN

China's Power Sector Reforms

Where to Next?

HD
9502
.C62
C4965
2006

10082617

INTERNATIONAL ENERGY AGENCY

The International Energy Agency (IEA) is an autonomous body which was established in November 1974 within the framework of the Organisation for Economic Co-operation and Development (OECD) to implement an international energy programme.

It carries out a comprehensive programme of energy co-operation among twenty-six of the OECD's thirty member countries. The basic aims of the IEA are:

- To maintain and improve systems for coping with oil supply disruptions.
- To promote rational energy policies in a global context through co-operative relations with non-member countries, industry and international organisations.
- To operate a permanent information system on the international oil market.
- To improve the world's energy supply and demand structure by developing alternative energy sources and increasing the efficiency of energy use.
- To assist in the integration of environmental and energy policies.

The IEA member countries are: Australia, Austria, Belgium, Canada, the Czech Republic, Denmark, Finland, France, Germany, Greece, Hungary, Ireland, Italy, Japan, the Republic of Korea, Luxembourg, the Netherlands, New Zealand, Norway, Portugal, Spain, Sweden, Switzerland, Turkey, the United Kingdom and the United States. The European Commission takes part in the work of the IEA.

ORGANISATION FOR ECONOMIC CO-OPERATION AND DEVELOPMENT

The OECD is a unique forum where the governments of thirty democracies work together to address the economic, social and environmental challenges of globalisation. The OECD is also at the forefront of efforts to understand and to help governments respond to new developments and concerns, such as corporate governance, the information economy and the challenges of an ageing population. The Organisation provides a setting where governments can compare policy experiences, seek answers to common problems, identify good practice and work to co-ordinate domestic and international policies.

The OECD member countries are: Australia, Austria, Belgium, Canada, the Czech Republic, Denmark, Finland, France, Germany, Greece, Hungary, Iceland, Ireland, Italy, Japan, Korea, Luxembourg, Mexico, the Netherlands, New Zealand, Norway, Poland, Portugal, the Slovak Republic, Spain, Sweden, Switzerland, Turkey, the United Kingdom and the United States. The European Commission takes part in the work of the OECD.

© OECD/IEA, 2006

No reproduction, copy, transmission or translation of this publication may be made without written permission. Applications should be sent to:

International Energy Agency (IEA), Head of Publications Service,
9 rue de la Fédération, 75739 Paris Cedex 15, France.

FOREWORD

At the International Energy Agency (IEA), we believe that access to modern energy services is essential for the social and economic development of every country and, more broadly, of the global system. Without reliable and affordable electricity, children have no light to study, food spoils, medical care cannot be provided, and the motors that drive industrial productivity remain idle. Making this service available, however, can be difficult – especially in a vast nation like China where more than 1.2 billion people live, dispersed over 3.7 million square miles of land covering mountains, deserts and remote rural areas.

China's rapid pace of economic growth has created a strong appetite for electricity. In the last two years alone, the country has added nearly 117 GW of capacity – approximately equal to the total electricity capacity of France or Canada. No other country has been able to mobilise its resources to achieve such astounding expansion, particularly after initiating reform and unbundling its power sector. The government of China should be commended for this impressive feat.

Despite this notable progress, challenges remain. China must be able to balance the pressures of increasing electricity demand with growing concerns about energy security and environmental impact. Its regulatory framework needs to be designed to ensure investment, encourage energy efficiency, minimise cost and reduce emissions – a very tall order in any circumstance! A number of IEA countries have developed energy policies in pursuit of similar goals. Their results have been mixed, but many lessons have been learned.

This book aims to draw insights from IEA countries' experiences that may be useful for policy makers formulating China's next steps in power sector reform. At the same time, IEA countries can benefit from this analysis of China's experience in building one of the world's largest power sectors. In this increasingly global society, the more we learn from each other, the better we can prepare for a sustainable energy future.

Claude Mandil
IEA Executive Director

ACKNOWLEDGEMENTS

This book has been prepared with the help of a number of people both within China and from outside, including international experts on China's energy sector and Chinese government officials, as well as members of the IEA Secretariat.

The study was launched on an IEA mission and workshop to discuss the challenges facing China's power sector, hosted by the National Development and Reform Commission (NDRC) in Beijing (November 2004). A second IEA mission to China (February 2006) was an opportunity for further discussion of emerging conclusions. The IEA would like to thank the following Chinese institutions for the time and effort given to help our understanding of the issues through these missions: the NDRC; the State Electricity Regulatory Commission (SERC); the State Grid Corporation (SG); and the State Power Economic Research Centre (SPEC).

The main author of this book is Caroline Varley, consultant on power sector reform and formerly head of the IEA's Energy Diversification Division. Jeffrey Logan (now with WRI, formerly the IEA's China Programme Manager) started the process of developing the book, which was taken over by Jonathan Sinton, China Programme Manager in the IEA Secretariat's Office of Non-Member Countries, who steered the book to its conclusion and guided the development of its key messages. Bill Ramsay and Yo Osumi encouraged and directed the work. Others in the IEA Secretariat also provided valuable input: Noé van Hulst, Ulrik Stridbaek, Jens Laustsen, Hideshi Emoto, Paul Waide, Brian Ricketts, Emmanuel Bergasse, and Nancy Turck.

Thanks also go to the following international experts for their contributions and support: Philip Andrews-Speed and Ma Xin (Centre for Energy, Petroleum and Mineral Law and Policy, University of Dundee, United Kingdom); Wei Bin, Ge Zhengxiang, and Jiang Liping (SPEC); David Moskovitz and Richard Cowart (The Regulatory Assistance Project); Wang Wanxing (The Energy Foundation); Doug Cooke (Department of Industry, Tourism and Resources, Australian Capital Territory Government); Zhao Jianping (World Bank, Beijing Office); Per Christer Lund (Counsellor, Science and Technology, Royal Norwegian Embassy, Tokyo); Lin Jiang (LBNL); Xavier Chen (BP China); and Hans Nilsson (Chair, IEA Implementing Agreement on DSM, and CEO, FourFact consultancy).

The IEA staff who worked on the production of the report also deserves thanks: Loretta Ravera, Muriel Custodio, Rebecca Gaghen, Bertrand Sadin, and Sylvie Stephan. Marilyn Smith edited the report. Chantal Boutry provided administrative support during the course of the study.

Any questions or comments should be directed to Jonathan Sinton, China Programme Manager, Jonathan.Sinton@iea.org.

TABLE OF CONTENTS

INTRODUCTION

The main objectives of this report are to identify next steps in the further reform of China's power sector over the next 2-5 years, and to provide, for the government's consideration, a set of practical recommendations in support of China's strategic goals to boost economic growth and reduce energy intensity. China has taken important and courageous steps to reform its power sector over the last two decades. In some areas, it has gone further than many other reforming countries, including some in the OECD. But the reform process needs to be reinvigorated. It is a dangerous illusion that China's recurrent supply problem is solved because capacity looks reasonably comfortable once again, as this report goes to print.

For such a large country as China, which is engaged in a complex transition toward a socialist market economy, it would be impossible to cover all the issues related to the power sector. In particular, this report does not examine the issue of fuels and technologies for power generation. It does, however, emphasise the fact that coal, which currently accounts for some three-quarters of generation, can be expected to remain dominant for the foreseeable future. The emerging regulatory framework for the development of competitive power markets should take account of this, and seek to mitigate the harmful environmental effects of coal-powered generation.

This report's recommendations are based on the principle that competitive power markets remain the best long-term goal – a goal reiterated in China's recently issued 11th Five Year Plan. But in its progression towards a competitive market, China's priorities at this stage should be to strengthen the institutional and governance framework, to review actions for tackling coal pollution, and to develop and implement specific reforms for more cost-reflective and efficient pricing, which will provide incentives for investment in energy efficiency and for strengthening the grid and generation. Actions are also needed to reinforce the groundwork for competitive markets: China should consider the development of some competitive trading, initially on a modest basis, across its regions and provinces. China needs to review and reaffirm its strategy for power sector reform, and to ensure that there are strong mechanisms for implementing further reforms. A key lesson from other reform experiences is that strategic reform goals need to be clearly articulated, and that how reform will be implemented is as important as what reforms are needed.

Another major lesson from reform experiences is the opportunity for China to leapfrog other reformed jurisdictions by integrating energy efficiency and environmental goals into its regulatory framework for competitive power markets from the start. Managing demand, as well as improving the supply infrastructure, is the best way for China to meet its strategic goals.

Efforts have been made to review the most relevant parts of the considerable work which had already been carried out on China's power sector reforms in the course of developing this report. Where observations and recommendations still appear relevant or need to be underlined, they are highlighted[1].

This report is divided into three parts. This first part summarises the main body of this report, draws out key messages and sets out recommendations. The second part reviews the main features of China's power sector, the current governance and regulatory framework, and assesses performance. The third part analyses the issues that need action in the near term, ending with a section on considerations for the longer term.

1. The World Bank and the Energy Foundation, in co-operation with numerous Chinese experts, have been especially active over the last decade. See, for example, Shao *et al.* (1997), World Bank and Energy Foundation (2000), Energy Foundation (2002), Hu *et al.* (2005), and numerous other works cited in this report.

KEY MESSAGES

China has set for itself two formidable strategic goals: doubling 2000 GDP by 2010 and reducing energy intensity by 20% over the next five years. The next steps in power sector reforms should help to achieve these objectives. Following careful analysis, the International Energy Agency offers the following key messages:

- The development of fully competitive power markets should remain the long-term goal. In the progression toward this goal, near-term priorities should be:
  - To strengthen the institutional and governance framework.
  - To review actions for tackling coal pollution.
  - To develop and implement specific reforms for more cost-reflective, efficient pricing and investment, providing incentives for investment in energy efficiency and strengthening the grid and generation.

- Near-term priorities should also include actions to lay a stronger foundation for the evolution of competitive markets across the country, and a first set of measures to stimulate basic competitive trading across China's regions.

- China needs to review and reaffirm its strategy for power sector reform, and to ensure that there are strong mechanisms for implementation of further reforms.

- Greater transparency is the key that will help to unblock further reform progress across all fronts. This includes improving data collection and analysis on the power sector so as to improve understanding of supply and demand developments.

- China has the opportunity to leapfrog other reformed jurisdictions by integrating, from the start, energy efficiency and environmental goals into its regulatory framework for competitive power markets.

EXECUTIVE SUMMARY

ELECTRICITY POWERS ECONOMIC DEVELOPMENT

China's economic development has been spectacular. Two decades of sustained economic growth at an average rate of 9.5% per year have resulted in a six-fold rise in China's GDP. More than 200 million people have been lifted out of poverty, the best performance by any single country in recorded history. China's economy is now larger than those of a number of major European countries; in five years, it may be exceeded by only three OECD countries. The rapid pace of economic change is likely to be sustained for some time.

It is increasingly clear that a reliable power sector is important to support economic development. In a very broad sense, China has succeeded in supplying the electricity needed to drive economic growth and raise living standards. By the same token, chronic supply shortages have the potential to undermine sustained future economic growth.

Impressive as these achievements are, China needs to go even further if it is to meet the development levels of the more advanced countries – which China seems prepared to do. Under its 11th Five Year Plan, China outlines formidable strategic objectives (Box 1) that include doubling 2000 per capita GDP by 2010 and reducing energy intensity of GDP by 20% over the next five years. This requires major new investments aimed at delivering more reliable and less polluting power, and at reversing the current rise in energy intensity trends. Reducing or, at the very least, containing rising pollution is another major goal.

Box 1 Electricity in the 11th Five Year Plan

In March 2006, China released its outline of the 11th Five Year Plan, intended to guide socio-economic development of the country from 2006 to 2010. The new Plan, like previous ones, will be the touchstone for all manner of government policy at the national and local levels for the next several years. A number of its 48 passages deal with electricity, including exhortations to proceed with power sector reform, and to develop and deploy more efficient and environmentally friendly systems for generation and transmission. Several examples follow:

Chapter 6. Transform the face of the countryside

"Vigorously develop rural biogas, crop residue fired electricity, small hydropower, solar energy, wind energy and other renewable energy, and improve the rural power grid."

Chapter 9. Deepen rural reforms
New key projects in rural construction
"Village Electrification and Green Energy County Project: Establish 50 green energy demonstration counties, using grid extension, wind power, small hydropower, solar photovoltaic power, etc. to bring electricity to 3.5 million homes now without power."

Chapter 19. Implement the overall strategy of regional development
In western regions: "Build electric power bases and projects to send western power eastwards."
In central regions: "Develop mine-mouth power plants and combine coal-electricity operations."

Chapter 24. Enlarge the scope of environmental protection
"Accelerate the installation of flue gas desulphurisation (FGD) equipment at existing power plants, require new power plants to install FGD to meet emissions requirements... Make 90% of existing power plants meet emissions standards."

Chapter 31. Maintain and perfect the basic economic system
"Deepen power sector system reform, consolidate the separation of generation and grid, accelerate the separation of parents and subsidiaries, steadily advance the separation of transmission and distribution, and establish regional power markets."

Chapter 34. Perfect the modern market system
"Advance electricity price reform, gradually set up a system with competitive markets for generation and retail power, while government sets prices for transmission and distribution."

CHALLENGES FOR THE POWER SECTOR IN CHINA

In reality, China cannot achieve its ambitious goals without further power sector reforms; current policies cannot sustain growing demand. Despite important reforms and significant investment over the last two decades, China's power sector still grapples with reliability issues and a "boom/bust" supply cycle that fluctuates regularly between periods of highly disruptive supply shortage and inefficient overcapacity. At the same time, pollution from the power sector is poorly restrained and is expected to continue rising against a background of rapidly increasing demand. China is now the world's second largest electricity consumer; the power sector is the country's largest polluter.

Specifically, China's challenges in the electricity sector relate to four key areas:

- **Supply/demand imbalances.** China's boom/bust cycle is exacerbated by significant gaps between power generation and consumption in many provinces (Figure 1). For example, chronic supply shortages afflicted the country between 2002 and 2005.

As a result, 25 of China's 31 provinces and major municipalities sustained significant power losses. Industry experienced enforced closures and consequent losses; households felt the impact of significant reductions in basic comfort levels. China appears to be moving back into the boom part of its cycle, with some overcapacity expected for 2007. With such a fast growing economy, this type of supply/demand pattern might be expected. The challenge is how best to minimise these fluctuations.

- **Rising energy intensity.** Following a decline lasting nearly two decades, intensity started to rise again about four years ago. There are some exceptions, but levels of energy use efficiency in various sectors are often 20-40% lower than in developed countries. There is large scope for improvement.

- **Rising pollution levels.** China is home to five of the ten most polluted cities in the world. Acid rain falls on one-third of China's territory and one-third of the urban population breathes heavily polluted air. Poor air quality imposes a welfare cost of between 3-8% of GDP. China's power sector is the single largest culprit, responsible for an estimated 44% of SO_2 emissions, 80% of NO_X emissions, and 26% of CO_2 emissions. While per capita greenhouse gas emissions are still low, the power sector is now China's largest source of these emissions.

- **Need for additional investment.** More than 10 million rural Chinese still have no access to electricity and China's electricity sector still relies heavily on public funds. More investment is needed, from different sources, and must be directed to specific issues: cleaner generation; transmission and distribution; and energy-efficiency measures on the supply and demand sides. So far, reforms have not created incentives for investments in energy efficiency. Some estimates suggest that USD 50-70 billion per year may need to be invested in generation. This is approximately double the current rate (at least as regards the grid) and has never been achieved in the past.

AN OVERVIEW OF REFORMS TO DATE

China has made considerable progress in reforming the structure, governance and institutional framework of its power sector. This has been a long, drawn-out process of transition that began in the 1980s within the context of wider economic reforms to promote growth and economic development. The rate of reform in the power sector has been slower than in most other industries but significant changes have taken place and now provide the foundation for further development.

By the end of 2002, China had moved from a single, vertically integrated utility to two grid companies (a large one covering most of the country, and a small one in the south) and a diverse set of generation companies (five large companies that were spun off the original incumbent and a large number of other companies). This provides a potential basis for developing vigorous competition, once the new industry structure is supported by effective governance and regulation. In addition, China launched competitive power markets on a trial basis in three regions (more are

Figure 1 Patterns of electricity generation and consumption in China, 2003

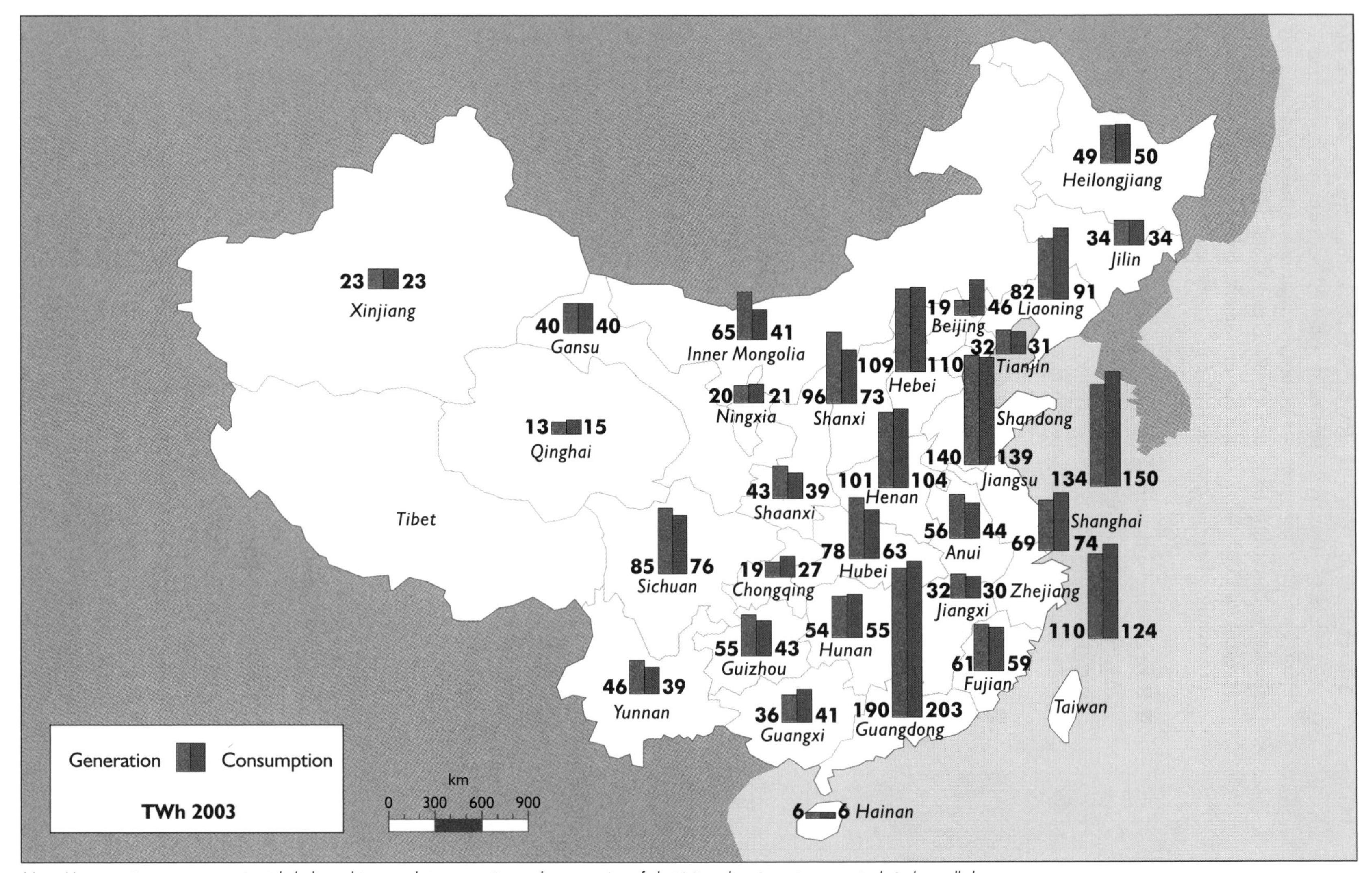

Note: Many provinces are approximately balanced in regards to generation and consumption of electricity; others import or export relatively small shares.
Source: Editorial Board of the China Electricity Yearbook (2004).

planned) and took a first step towards independent regulation by establishing the State Electricity Regulatory Commission (SERC).

One of the remaining challenges in these initial reforms is that ownership of the power sector is still largely with the state. The grid is mostly owned by the central government, and generation companies are often linked, directly or indirectly, with local state interests. Further reform proposals exist, particularly as regards pricing, but have not yet been implemented. The 11th Five Year Plan calls for expanding electricity structural and price reforms but does not provide the details of specific measures and timetables.

WHERE TO NEXT?

This report's recommendations, like the 11th Five Year Plan's objectives for power sector reform, are based on the principle that competitive power markets are the best long-term goal for China. However, competitive power markets are not an end in themselves; rather they are a means to an end: access to environmentally sustainable electricity services to achieve China's social and economic welfare objectives. To serve as an effective instrument, many electricity policies must be considered simultaneously: regulatory policies and structures must integrate competition principles and cost-reflective, competition-based pricing alongside policies to encourage energy efficiency and policies for the environment. Without a holistic approach, competitive markets can raise problems for demand management (*e.g.* dispersing incentives to reduce demand) and the environment (*e.g.* because environmental costs and benefits are not yet appropriately reflected in power pricing and investment decisions, system dispatch sometimes favours dirtier plants). China's progress towards competition should proceed carefully. Important actions should be taken now to improve economic and energy efficiency without compromising the long-term goal, and to lay a sound basis for a fully competitive market in due course.

CHINA NEEDS TO REAFFIRM A CLEAR STRATEGY FOR ITS POWER SECTOR

The objectives and tasks defined in the State Council's 2002 policy document remain the government's formal baseline. The document makes it clear that the strategic goal is to develop a competitive, market-based, power sector, as a means to ensure an efficient and reliable power supply, and to protect the environment. Important parts of this strategy have been implemented decisively and effectively. This applies in particular to the separation of the grid from generation, which has allowed the emergence of a large number of generators, and the judicious split of the grid into two main companies. The establishment of a regulator (SERC) has underlined a commitment, at least in principle, to the importance of independent regulation. The considerable efforts that have been put in to establish pilot regional power markets with a view to gradual build up of competition in these markets is another very positive indication of steady commitment to developing a long-term competitive market and reaping the benefits of improved efficiency and reliability. China's achievements to date, particularly the disaggregation of the grid and generation, provide a stark contrast to many other OECD countries that have been slower to take determined action.

Some aspects of the original plan, however, have yet to be implemented, and those elements that have advanced need more work. At present, several factors undermine confidence in China's commitment to reform. Pricing reforms aimed at supporting more competitive markets and introducing greater economic efficiency appear to have stalled. Environmental issues have been approached in a fragmented way, with insufficient effort to integrate environmental considerations into the emerging framework for economic regulation. Structural unbundling has begun, but needs to be completed with a clean separation of generators from the grid. SERC is not yet the fully independent and empowered regulator required to oversee the new market structures. In addition, developing pilot competitive power markets appears difficult; they only cover a small portion of the markets in a few parts of the country, and there is little evidence of a start to competition elsewhere in the country. Significant delays in implementation of key reforms, such as pricing, suggest uncertainty as to next steps – and possibly raise questions about the current strategic thrust of the reform process.

The combined effect of these delays and incomplete developments is quite serious. China is caught between the old planning mechanisms and a new approach. Much of the power sector remains trapped in a governance system that consists of an uneasy mix of socialist style planning and more market-based regulation. The transitional governance and institutional systems, and the delayed development of more efficient pricing, could mean that the big challenges afflicting the sector – reliability problems, growing energy intensity, pollution, and inefficient investment – could even get worse.

It is important, therefore, that China reaffirms its commitment to a clearly articulated strategy for reform. This would boost confidence among stakeholders (including, not least, investors) that there is a plan that will bring greater clarity to the market environment, as well as continued commitment to implementation and improvement. One of the most striking lessons from reform processes around the world (not just in the power sector) is the frequent failure to define the end point of the reform process and clarify the strategy for arriving at it. Reaffirming a clear strategy is also an opportunity for China to review – and to improve upon – the 2002 strategy. The most important objectives of the review would be the following:

- **Integrate energy-efficiency goals and measures to achieve these alongside existing demand-side goals.** In order to align China's power sector reforms (both now and for the longer term) with the country's overall development strategy, policies to reduce demand should be given as much consideration as those aimed at strengthening investment in generation and the grid. A two-pronged approach has the best prospects for mitigating (if not eliminating) China's uncomfortable boom/bust cycle of supply/demand imbalances.

- **Ensure that the emerging regulatory framework includes incentives to improve environmental performance.** Actions taken so far appear to segregate environmental policies from the economic regulation of the power sector, limiting the overall effectiveness of regulation. The new regulatory framework must give all participants a clear view of costs, including environmental costs. Incentives to produce cleaner power based on the "polluter pays" principle (taxes, subsidies, obligations or quotas, etc.) must be underpinned by better collection and communication of emissions data alongside other information.

- **Distinguish between near term and longer term actions to improve performance.** China needs to devote effort now to reform activities that can yield positive near term benefits while also helping to lay the groundwork for fully competitive markets. These include: strengthening the institutional framework; integrating energy efficiency and environmental objectives more firmly into current regulation and future reform plans; and implementing pricing reforms to support improved economic and energy efficiency. Taking actions, including modest steps towards competitive trading, to establish a sounder groundwork for the efficient development of fully competitive power markets in due course is also important at this stage.

GREATER TRANSPARENCY KEY TO REFORM STRATEGY

China remains distant from the levels of transparency needed for an effective reform process and for the efficient operation of its power markets. Because relevant information is not always made public, it is hard to assess current levels of profit, subsidies and cross subsidisation. It is also difficult to determine the amount of public funds flowing into the power sector, especially for infrastructure development. Moreover, it is not clear whether stakeholders have opportunities to participate in the regulatory process.

Increased transparency can be integral to unblocking further reform progress across all fronts. It can improve the quality of the regulatory process and the efficiency of new power markets, and may also smooth the relationships between the centre and local levels of government. One essential element of the framework for transparency is more open exchange of information, particularly reliable data on the power sector and supply/demand developments. China already collects and publishes a great deal of information, but more, and more timely information will help lay the foundations for better performance across all classes of generators and consumers.

A REFORM CHAMPION AND BROADER LEADERSHIP NEEDED TO SUSTAIN MOMENTUM

Reform processes elsewhere point clearly to the importance of having a strong and stable policy reform advocate (or advocates) in the central government. This helps to minimise uncertainty, maintain coherence and ensure stability, thereby building vitally important confidence among market players. In the current absence of a comprehensive energy ministry, the NDRC is the main focal point for the China strategy, with SERC contributing in significant ways. Yet these two agencies have different mandates and oversee different elements of the strategy.

The power sector raises many complex issues that are best dealt with by specialists. It also requires integration with broader policy and tight co-ordination with other areas of government (finance, environment, rural development, etc.) that play a role in power sector reform. Thus, there is a clear need to develop a new comprehensive energy ministry or agency with a mandate to oversee the power sector within the broader framework of the entire energy sector. In addition, it may be appropriate to utilise the Chinese approach of setting up a "Leading Group". By bringing together a widely based group of stakeholders, the Leading Group could help to sustain coherence, identify essential activities and improve understanding of often complex issues that are specific to the power sector. It might also help to prevent the "hijack" of the reform debate by minority vested interests.

FIVE AREAS FOR ACTION IN THE NEAR TERM

China confronts a tremendous array of choices in its next steps for reforming the power sector. It may be helpful to sort them into five areas: strengthening institutional capacities, promoting environmental goals, making pricing and investment more efficient, promoting energy efficiency, and preparing for competitive markets.

Strengthening institutional capacities

As a result of the recent reforms, China's power industry is now governed by three key institutions: the National Development and Reform Commission (NDRC); the State Electricity Regulatory Commission (SERC); and the State-owned Asset Supervision and Administration Commission (SASAC). The establishment of SERC (2002) was a clear signal of China's commitment to independent regulation in the power sector. Despite State Council efforts to strengthen its powers, SERC remains quite weak, in terms of both powers and resources. In addition, its independence from the NDRC is not clear; policy and regulatory functions are muddled between the two entities. Ultimately, SERC's independence remains questionable *vis-à-vis* the government and the regulated companies.

To date, China's reforms have focused primarily on disaggregating the industry from government functions. The challenge, at this stage, lies in developing a robust framework within government – and across all levels – to regulate this "new" industry. China's accession to the WTO in 2001 has accelerated change towards a socialist market economy, but it is nevertheless a slow process.

In fact, several factors underline the need for caution in moving ahead too quickly towards full competition. Power sector reform in China faces some important governance challenges that cannot all be resolved by policies aimed at the sector itself. Awareness of the weak spots in the broader framework, as identified above, may prompt the development of strategies to better manage them. It may be possible

– and even desirable – to strengthen capacities in some areas (*e.g.* monitoring anti-competitive behaviour in power markets) ahead of broader governance progress (*e.g.* development of a competition authority).

Significant institutional improvements are needed to reinforce the prospects for further effective reform. China should take specific steps to:

- Complete revisions to the Electricity Law to frame and support further reform.
- Further separate regulatory from policy functions in NDRC and SERC.
- Further empower SERC, *i.e.* if it cannot be given responsibility for pricing, it should at least make public recommendations that can be publicly debated.
- Ensure that SERC has the capacity to monitor and manage competition issues in the market (until such time as a competition authority is established).
- Ensure that energy efficiency and environmental goals are incorporated into policy decision-making processes and carried through in the regulatory framework that will implement the policies.
- Strengthen SERC's staff and resources, including increased expertise on energy efficiency and environmental issues.
- Strengthen local regulatory capacities (*e.g.* tariff setting analysis capabilities).
- Improve data collection and analysis on the power sector.

Promoting environmental goals by tackling coal pollution

Regardless of origin, all forecasts, scenarios and plans concerning power generation in China point to decades more of coal's dominance. A major energy strategy study in 2004 found that, under varying assumptions, coal may continue to account for between 59% and 70% of generation capacity in 2020 (China Energy Development Strategy and Policy Research Group, 2004). For this reason the power sector is – and will remain – the largest emitter, by far, of some of the most important airborne pollutants in China.

An analysis of the model results for China prepared for the IEA's *World Energy Outlook 2004* shows that implementing stronger policies could significantly reduce growth in coal use and in the resulting carbon dioxide emissions. The IEA's alternative scenario for China shows that policies need to be deployed on both supply and demand sides. Much of the change would be driven by demand-side policies aimed at reducing growth in electricity consumption. The supply side offers various means of improving the environmental performance of power plants: changes in fuel; improvements in plant efficiency; and emissions controls. Given the difficulty in quickly reducing coal's share in the fuel mix, the last two carry the most immediate promise. Despite improvements with new plants, China does not yet meet worldwide average efficiency norms and there is scope for major improvement in emissions controls.

Policy reform and a new regulatory framework aimed at introducing competition create an opportunity to incorporate incentives for adopting more environmentally friendly options. Preferred options that avoid distorting market decisions and seek to minimise the effects on retail prices might include:

- Uniform generation performance standards or a pollution fee to increase the likelihood of cleaner plants being dispatched.

- Investment planning methodologies and licensing rules that encourage investment in cleaner plants via a proper reflection of environmental costs and benefits.

- Pricing that incorporates environmental costs across the value chain, including at the generation level. This would have significant impact on generation investment decisions and plant dispatch.

- Adapting the regulatory framework to assist in the enforcement of environmental regulations, and perhaps using the grid companies to help enforce environmental levies.

- Ensuring collection of necessary environmental performance data to enable tracking of emissions.

Enforcement of environmental regulations being a general weak spot, China should continue its efforts to increase transparency in this area, and consider formalising the channels through which information is made public and is acted upon (*e.g.* through public hearings). Pressure from well-informed public opinion already appears to be yielding positive results. In addition, China should review how the NDRC and SERC can best link into the work of the environmental agencies.

Policies must support more efficient pricing and investment

China's current pricing regulation needs to be understood in the broader context of a single buyer model of power transactions. Generators sell power to the grid companies at regulated prices, and the grid companies sell the power on to end users, whose prices are also regulated. There is no separate pricing for the grid; rather it is embedded in the other prices. The approach also applies in the recently established regional power markets, where end users are not yet allowed to transact directly with the generators. In short, the value chain is currently linked through the grid companies acting as single buyers.

The case for the further development of cost-reflective pricing to improve both the economic and energy efficiency of China's power sector is compelling and increasingly urgent. Despite periods of adequate supply, power shortages recur with uncomfortable regularity. As the sector struggles to keep up with rapid economic growth, this is expected to continue into the foreseeable future. More cost-reflective prices across the value chain would provide signals to trigger efficient investment and to curb consumption. Along with other incentives for improving energy efficiency, and more efficient and transparent investment planning, a more efficient pricing framework is a key mechanism for ensuring that supply can meet broad demand – at all times and over the long run.

Further reforms must address four crucial issues within the current framework. The first is lack of transparency in pricing regulation. The second is that cost-reflective pricing is not yet applied across the whole value chain, and that separate pricing for each service in the value chain has not yet been established. Third, current pricing does not include incentives for demand-side investment and end-use energy efficiency, nor for choosing the least environmentally damaging options. Finally, grid investment planning is not addressed in the most efficient way.

Addressing these issues requires near term action to reform pricing for generation, for the grid and for end users, as well as a thorough review of investment planning mechanisms. It does not require the implementation of fully competitive markets. These can come later, and pricing approaches can be adjusted during this later phase. In addition, pricing policy should begin to allow stronger cost pass-through down the value chain, encouraging energy efficiency and reducing pollution, through the following actions:

- A more transparent approach to pricing and the application of cost-reflective methodologies is needed to identify the extent of the use of public funds in the power sector – and to wean the power sector from this dependency. Creating a system that pays its own way is an essential foundation for effective competition.

- A more efficient pricing approach for generation, based on deployment of the proposed two-part (capacity and energy) pricing principle, as China appears to have in mind, will ensure that dispatch is based on each plant's marginal cost.

- Proposed plans for separate grid pricing and transmission pricing must be taken forward. The current absence of separate grid pricing distorts generation pricing and hinders efficient and adequate grid investments. Moreover, there are no mechanisms for players in emerging power markets to establish efficient trade based on the separate costs of power and of using the grid to dispatch the power.

- The process for grid planning and investment also needs review. The first stage should focus on introducing transparency and cost-reflectiveness so that infrastructure investments are more efficient. As competitive markets and competition itself develops, the process can evolve further so that investment decisions are guided by the market, under the oversight of the grid/system operator and regulator.

- China's end-user pricing regulation already includes some very positive features, such as time-of-use pricing. Although prices have been allowed to rise, residential consumers, heavy industry and agriculture continue to enjoy subsidised prices. There are no clearly articulated principles – such as cost-reflectiveness – to guide adjustments to end-user pricing. Methodologies for more cost-reflective pricing need to be established, alongside incentive and penalty schemes that encourage consumers to improve their energy efficiency.

- The process of unwinding subsidies should be started by designing lifeline support programmes for those who need it, to address the social effects of reform.

Policies for energy efficiency will help to manage demand

China's traditional approach to addressing power shortages has been to increase supply-side investment in generation and the grid. Yet energy is not an economic output that must be maximised at all costs. Rather, it is an input to the generation of goods and services such as heating, lighting, mobility, industrial products and consumer goods. Reducing the input needed to provide these goods and services would have benefits that reverberate throughout China's economy, including improved environmental quality, economic competitiveness and energy security (Energy Foundation, 2003). Furthermore, it is cheaper to reduce energy consumption than to increase supply (Figure 2).

Figure 2

Investing in end-use efficiency *vs.* demand-side capacity provides net gain

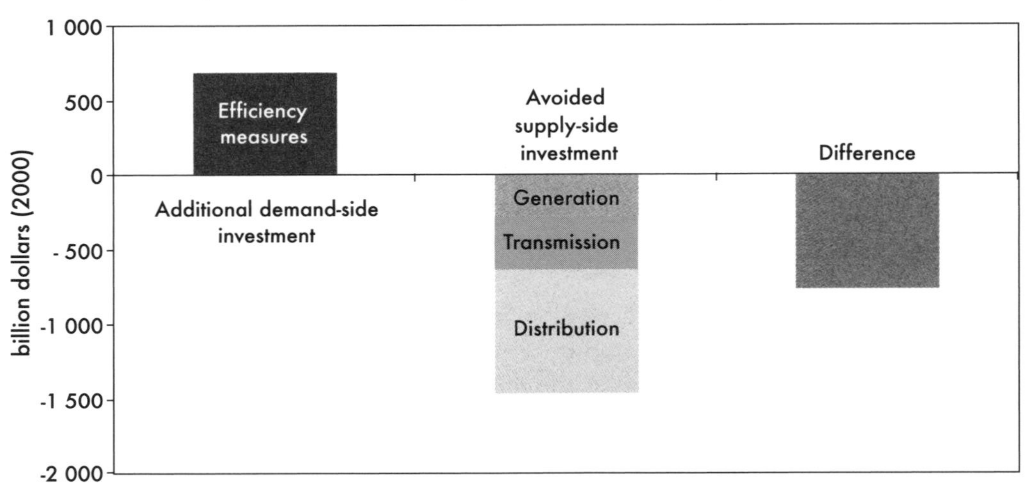

NB: This chart shows the cumulative difference in global electricity investment between the Alternative and Reference scenario from 2003 to 2030.
Source: analysis based on IEA (2004).

China has a tremendous opportunity to leapfrog other countries in developing a regulatory framework that takes account of the need for incentives for energy efficiency. Most countries that have reformed their power sectors were unprepared for how unbundling and competition affect traditional approaches to promoting energy efficiency. Sustaining funding and incentives for energy-efficiency programmes has become a major issue for them.

To date, China has relied too heavily on load shape management, to the detriment of policies aimed at reducing load. There is scope for deploying a more systematic approach to load shape management, which anticipates the peaks and valleys of demand. However, the main efforts at this stage should focus on reducing load over the longer term.

China has a long history of policies and programmes to promote energy efficiency, from direct support for investments to consumer education. Still, its electricity regulatory framework is not well-adapted to supporting energy-efficiency improvements that reduce demand over the longer term. The main focus of reforms so far has been on the supply side – *e.g.* increasing generation and grid capacity – but much less on saving energy. The reforms still lack a broad and sustained commitment to energy efficiency through demand-side management (DSM) and demand participation (*i.e.* responsiveness of customers through operation of a price mechanism). There is a major opportunity to integrate DSM and demand participation into the regulatory framework for competitive markets and into investment planning. This should, at the least, remove barriers and disincentives to energy efficiency and demand response. Going further, efficiency programmes could be strengthened by encouraging power companies to make it their business to provide not just electricity, but electricity services. In the near term, adjustments to the regulatory framework to promote efficiency could include:

- Developing sources of finance for investing in energy efficiency.
- Establishing incentives as well as mandatory requirements, for investing in energy efficiency.
- Promoting energy-efficiency "aggregators" to counter fragmentation.
- Initiating specific DSM activities covering all major fields of consumption, building on what China is already doing in areas such as lighting, appliances, and motors systems.

China should also consider participating in the IEA's Implementing Agreement on DSM, which already brings together 17 countries and the European Commission to share technology information, policy experience, data, and best practices[2].

Building the foundation for effective competitive markets

In the near term, there are two important objectives for reform: strengthening the framework for competitive trading and the careful introduction of limited competitive trading between different regions and provinces. There is scope to squeeze more capacity out of current infrastructure – *i.e.* to increase its efficiency – without necessarily having to build more. As regards the framework for competitive trading, grid independence is essential to reassure the market that system dispatch will be fair. If an incumbent utility retains control of the grid, or the grid company retains an interest in generation, it can easily limit or even exclude access to the grid by competing generators. System operation should also be fully independent of the competitive part of the market. China currently has a single buyer administered system, under which generators sell to the grid companies, not to final customers. Initiating, even in a modest way, wholesale country-wide competition, will help to lay the foundations for more extensive competition in due course. It will also strengthen near term efficiency, which is hard to maximise under the current single buyer approach.

Careful handling is required to ensure that relevant actions take account of continued state domination of China's power sector, at different levels. Near term actions should include:

2. For more information see http://dsm.iea.org/.

- Strengthening the independence of the grid and the fairness and efficiency of plant dispatch. This involves completing the separation of generators from the grid companies, and improving grid and system operation corporate governance.

- Developing more independent and well-governed generation companies. Detaching generating interests from local economic interests is a priority. Securing independence involves actions such as: establishing and monitoring "fit for purpose" regulatory accounts; developing clearer and more robust corporate governance rules and controls; and ensuring a level playing field for private interests.

- Moving towards competitive power trading, alongside and beyond the existing pilot markets. Trade between jurisdictions has often proved to be an effective lever for the development of competition and improved efficiency. Modest steps in this direction, and away from the current single buyer approach, could include: allowing the grid companies to establish their own power sale transactions with large consumers, independent of government directions; and, later, allowing generators to transact directly with large consumers.

- Strengthening market infrastructure for competitive trading by spreading private ownership, including foreign ownership, as well as ensuring that generation licensing does not obstruct market entry.

- Ensuring that there is a level playing field among all technologies and all fuels, as well as among demand-side resources.

- Policing anti-competitive behaviour through current institutions (SERC) while developing market rules that could later be enforced by a competition authority.

- Ensuring common basic rules across the regions to avoid future market fragmentation.

- Strengthening system security in anticipation of increased trade.

RECOMMENDATIONS

The following pages attempt to distil the preceding discussion on strategy and the five near-term priority areas into a concise guide for action, which is as specific as possible within the limits of this type of study. Each recommendation reflects practical steps that China can take now. Some of these recommendations are not new. However, among the many actions open to China, they are the most urgent for building a healthier power sector to serve the country's many needs – both now and in the future. Several recommendations could become the objects of focused technical assistance, should China be interested.

Each recommendation is elaborated in greater detail in the body of this report, within the corresponding chapters indicated by the headings.

REAFFIRMING A CLEAR STRATEGY

Reaffirm long-term strategy, undertake near-term actions

China should reaffirm and update its strategy for further power sector reforms, thereby reiterating its commitment to the long-term development of competitive power markets. The strategy should explicitly integrate energy-efficiency goals and environmental objectives. The strategy should distinguish between near-term and longer-term issues, and specify related actions. Near-term priorities should include efforts to:

- Strengthen the institutional framework.
- Integrate energy efficiency and environmental objectives into current regulation and future reform plans.
- Implement pricing reforms to support improved economic and energy efficiency.
- Establish a solid foundation for developing competitive power markets, including a first set of measures to stimulate basic competitive trading across China's regions.

Update electricity law

China should move swiftly to update its electricity law to provide a firm foundation for further reforms and to clearly establish the strategic objectives for the power sector. This is an opportunity to emphasise a new holistic approach for the regulatory framework that:

- Incorporates energy efficiency and environmental considerations, as well as economic regulation of the sector.
- Confirms SERC as an independent regulator.
- Allows for adaptation as the power sector evolves, to avoid the need for further revisions in the near future.

Appoint a reform champion

China should identify a focal point within the institutional structure that will act as the "champion" for taking policy reforms forward. The champion can play a key role in clarifying the separation between policy and regulatory functions; it needs to be adequately resourced to:

- Provide further policy support for reform.
- Establish a mechanism to reactivate momentum for the reform process within government.
- Create a "Leading Group" to tighten co-ordination on electricity issues between different parts of government.

Without losing sight of the main overall strategy, the champion should be prepared to carry out many specific tasks, such as:

- Dealing with vested interests.
- Promoting reform "ownership".
- Adjusting the policy programme to fit developing circumstances.
- Encouraging consumer participation in the reform process.
- Establishing an implementation plan, and ensuring that it is met.
- Encouraging new approaches to compliance and enforcement.
- Establishing a mechanism for monitoring and evaluation.

Increase transparency, improve communication

China should take steps to increase transparency of the reform process, and should improve communication of its strategy and specific reform plans. Relevant information should be publicised and disseminated to appropriate target audiences:

- Opportunities should be created for stakeholders to participate in the reform process, for example by commenting on specific proposals.

- The broad picture, objectives and expected results should be explained to users and the wider public.

- More technical information should be made available to market players.

Improve data collection, analysis and access

China should find ways to improve its systems for data collection and analysis on the power sector, and provide stakeholders with better access to information. This is an essential element of transparency and will increase stakeholder understanding of power sector developments. SERC should play a leading role in this process, in order to avoid potential distortions by parties with a vested interest. The institutional design should ensure long-term support for collection of statistics that cover all relevant aspects of the electricity system. Key information that needs to be available includes:

- Current demand and demand peaks.

- Projections of future supply and demand by region.

- Clear and detailed knowledge of the grid and generation infrastructure capacities.

STRENGTHENING INSTITUTIONAL CAPACITIES

Appoint SERC as independent regulator

China must take action to strengthen SERC's capacities so that it can evolve as an independent regulator. This includes ensuring that SERC is empowered to take the necessary regulatory actions in the market as it now stands, without waiting for further developments in competitive power markets. These actions should clarify SERC's objectives, responsibilities and powers – and strengthen its institutional features.

Clearly define SERC's mandate

SERC's objectives and responsibilities should be clarified and communicated to all stakeholders. In addition, SERC should be provided the enforcement powers needed to discharge these responsibilities effectively, across all of the following areas:

- Pricing.

- Regulatory oversight of market players (generators and the grid).

- Oversight of system dispatch and system security.

- Shaping the regulatory aspects of energy efficiency and environmental policy development.

- Data collection and analysis of the power sector.

Prepare SERC for enhanced role in pricing

SERC should gradually develop an enhanced role in pricing decisions, with the long-term goal of taking over this responsibility. In the near term, SERC should be formally empowered to make the advice it provides to NRDC easily available to broader publics. SERC should be given responsibility for monitoring costs and enforcing the pricing rules.

Empower SERC to address anti-competitive behaviour

Pending the establishment of a competition authority, SERC should develop its capacities for identifying and monitoring anti-competitive behaviour. Staff with the necessary competences should be recruited or trained, perhaps through exchanges with other jurisdictions that have successful track records in dealing with these issues.

Strengthen SERC's institutional features

In order to support the expansion of SERC's institutional role, it is critical to, in the first place, strengthen staff and resources. Effort should also be made to increase transparency, with the aim of creating a stronger public presence for SERC, both within the market and with the wider public.

Enhance regulatory capacity across government

Steps should be taken to enhance the regulatory capacities of local, as well as national, levels of government. As China has frequently done in other areas of reform, stronger provinces could be allowed to experiment with markets and market regulation, in order to provide test beds and to benchmark emerging best practices. Actions to support local development might include the transfer of experienced officials to poor areas, flexible pay scales, and, for the more senior posts, continued input from the central government on recruitment.

PROMOTING ENVIRONMENTAL GOALS: TACKLING COAL POLLUTION

Integrate environmental goals in policy reform

China should review and adjust its developing regulatory framework for competitive markets to ensure that it supports environmental goals. Areas that require particular attention include:

- Establishing fees or emissions standards to help secure the dispatch of cleaner plants.
- Incorporating environmental costs and benefits in power pricing.
- Conducting a review of investment planning methodologies and licensing rules to encourage cleaner investments.

Create mechanisms to enforce environmental regulation

China should seek to adapt its regulatory framework to support the enforcement of environmental regulations. This might include actions that help to track emissions and involve grid companies in enforcement activities.

Link environmental goals with market competition

China should review its institutional structures to ensure that they are capable of promoting the development of policies that ensure consideration of environmental goals as competition develops. NDRC and SERC should establish means of linking into the work of the environmental agencies. In addition, an institutionalised co-operation mechanism should be established between SERC and SEPA to take advantage of the complementary nature of these two agencies. SERC could be tasked with the specific responsibility of integrating environmental goals into the economic regulatory framework; SEPA could assess the environmental consequences of reform proposals for the power sector. Both agencies would need adequate staff resources and competences to do this; a formal co-operation agreement could create opportunities to, for example, provide for regular meetings and staff exchanges between the two agencies.

TOWARDS MORE EFFICIENT PRICING AND INVESTMENT

Create transparent pricing approach that reflects real costs

China needs to develop an overall approach to pricing across the entire value chain that is transparent and reflects the costs of electricity production and transportation to end consumers. Ultimately, this will create a power sector that pays its own way and no longer depends on public funding, which could be better deployed elsewhere. This recommendation is linked to others, notably the need to corporatise the generation sector into well-defined companies with clear ownership, responsibilities and objectives.

Implement two-part pricing

China should implement its proposed two-part pricing principle to provide the basis for more efficient plant dispatch, based on each plant's marginal cost (*i.e.* system dispatch should be based on the power price, with the cheapest plant being dispatched first).

Establish separate grid pricing

China should implement its proposal to establish separate pricing for the grid. However, it should also aim to move away from postage-stamp pricing in due course. The longer term goal should be to migrate towards a transmission-pricing system that: balances incentives for economic efficiency and investment; creates incentives for energy efficiency; balances development of the power system across the country; and promotes renewables. At the same time, China should improve methods for investment planning to better reflect real costs, and consider developing locational signals (via the auctioning of transmission capacity) as competitive markets develop.

Develop a cost-reflective approach to grid planning and investment

China should move away from its current "bottom-up" and non-cost reflective approach to grid planning. It should develop a transparent process for grid planning and investments that takes account of costs, as far as possible. In addition, the plan should ensure that energy-efficiency investments are properly considered as an alternative to supply-side investments.

Unwind subsidies, deploy new pricing, incentive and penalty mechanisms

To support cost-reflective pricing, China should start to unwind subsidies and cross-subsidies, and should increase the transparency of public funding for the power sector. These actions are steps toward eventually eliminating public funding. At the same time, it should continue to deploy time-of-use pricing and consider creating incentives and penalties that encourage consumers to improve their energy efficiency (inclining block prices, linking prices to efficiency standards for buildings via hook up fees, etc.).

Create lifeline support mechanism for poorer populations

In order to mitigate adverse social and distributional effects that often accompany tariff rebalancing, China should develop a lifeline support mechanism aimed at the poorer parts of the population. This lifeline must be carefully designed to be available only to those who really need it. Features of the lifeline should include effective targeting, positive net benefits, administrative simplicity, and transparency.

MANAGING DEMAND

Strengthen load shape management

China should seek to strengthen its approach to load shape management through a more systematic deployment of options that have the capacity to smooth out the peaks and valleys of demand.

Create legal framework for demand-side management

China should secure an appropriate legal framework for the comprehensive development of demand-side management (DSM) activities. The framework should include provisions for:

- Financing DSM activities.
- Promoting investment in energy efficiency through incentive-based regulation, as well as mandatory requirements.
- Establishing measures to promote energy-efficiency aggregators.
- Conducting a review of the institutional structures for promoting DSM and establishing these structures on a clear and stable basis.
- Integrating DSM and demand response into the regulatory framework for competitive markets and investment planning.

Participate in international DSM activities

It would be highly beneficial for China to join the IEA's Demand Side Management Implementing Agreement (IA). Current members of the DSM IA actively encourage China's participation, both to share IEA country experiences with China, and to develop a procedure for sharing experiences between China and other countries.

TOWARDS EFFECTIVE MARKETS: ACTIONS FOR THE NEAR TERM

Separate generation interests from grid companies

Plans to complete the full detachment of generation interests from the grid companies should be completed, as soon as possible.

Improve corporate governance

China should review the corporate governance framework for the State Grid Corporation and the China Southern Power Grid Company Limited, with the aim of minimising state interference in the management of each enterprise. In addition, China should consider implementing OECD recommendations on corporate governance (OECD, 2005a), tailoring them to the power sector. This includes activities such as:

- Creating and enhancing the role of boards in state-owned enterprises (SOEs).
- Improving recruitment and performance evaluation procedures for senior management.
- Strictly separating the government's exercise of its ownership in SOEs from its regulatory and other functions.
- Eliminating interference in SOE management.

Implement "regulatory" accounting

Under SERC's management, China should establish a framework by which generation companies could produce and monitor transparent "regulatory" accounts.

Unbundle generation and state accounts

China should take steps to effectively and transparently unbundle generation accounts from the accounts of other state interests to which they are currently attached. In addition, rules should be developed to secure a neutral framework for competition in the generation sector, particularly between private and publicly owned players.

Expand competition beyond pilot markets

China should consider taking some modest, first steps towards the development of competitive power trading outside the pilot markets that are already established. A first step might be to allow grid companies to establish their own transactions (not directed by government) with large consumers (this might best be carried forward once grid companies are fully separated from generation interests). Subsequently, direct transactions between generators and large consumers could be allowed.

Encourage private investment: domestic and foreign

China should strengthen the policy and regulatory framework to encourage independent domestic and foreign investment in power generation. As competitive markets develop, it should ensure that these independent power producers have the third-party grid access necessary for carrying out transactions.

Develop mechanisms to manage anti-competitive behaviour

In the absence of a competition authority, China should pay special attention to strengthening the regulatory framework for managing anti-competitive behaviour. It should take action for the rapid development, implementation and enforcement of market rules that promote transparency and a comprehensive flow of information on market operations. The rules should be included in an instrument with legal status, under SERC's supervision.

Build flexibility into system operation and market rules

China should bear in mind the future possibility of a unified, country-wide power market. Thus, in the development of system operation and market rules, it should avoid developing multiple regional frameworks that would be difficult to integrate at a later stage. Effort should be made to identify those elements that need to be common from the start: a uniform bidding platform and common basis for transactions; consistent wholesale and grid pricing concepts across regional/provincial boundaries; etc. At the same time, SERC should be empowered to approve proposed variations in order to ensure that there are no impediments to future market integration.

Strengthen system security

China should act now to strengthen its framework for system security – rather than waiting until increased trade makes this a more urgent issue. Elements that require attention include:

- The legal and regulatory framework.
- Security standards.
- Co-ordination, communication and information exchange.
- Investing in technology and people.
- Asset performance and maintenance.
- Vegetation management.

I. THE STARTING POINT

MAIN FEATURES OF CHINA'S POWER SECTOR

Market structure and ownership

As in many other countries prior to reform, China's power sector was historically organised as a single vertically integrated utility, exclusively owned and operated by the central government[3]. Over the past 20 years, three major successive reforms have significantly changed this original structure (Box 2). The objective of the most recent restructuring, which separates the grid from generation, was to set the scene for the development of competition between generators, with the aim of improving operational efficiency and lowering prices. Today there are a very large number of generating companies in the market, including the five companies that were unbundled from the grid. In addition, the grid itself was separated into companies.

Box 2 Reforms to China's power: structure and ownership

Reforms to the structure and ownership of China's power markets began in the mid-1980s and can generally be divided into three phases.

A first set of reforms in the mid-1980s opened up generation to investment by third parties outside central government – mainly provincial and local governments, but also some domestic and foreign companies. Power plants built and purchased by these investors now account for over half of total capacity. These new investors in generation are sometimes called independent power producers (IPPs), but the term is largely misleading in the Chinese context. Most remain intimately linked to government (*e.g.* are owned by sub-central governments) and so are not really independent. During this period, the central government maintained sole ownership of the grid.

A second set of reforms took place in 1997, when most of the assets of the Ministry of Power Industry, *i.e.* nearly all of the grid as well as some 40% of generation capacity, were transferred to the newly formed State Power Corporation (SP). This marked the first step toward separation of market and regulatory activities, at least on paper. It is interesting to note that the State Power Corporation was in place during one of the few periods in recent Chinese history in which power supply exceeded demand.

3. Andrews-Speed (2003) provides a comprehensive review of the evolution of the power sector and other energy sectors. See also Xu (2002).

The third and most recent restructuring took place in December 2002, when SP (which by then had 46% of the country's generating capacity and 90% of transmission capacity) was disintegrated and its assets redistributed to eleven new or regrouped state-owned enterprises (SOEs) (Figure 3):

- Two grid companies: State Grid Corporation of China (SG), covering 26 provinces and China Southern Power Grid Company Limited (CSG), covering five southern provinces. SG divided its territory among five regional subsidiaries. The grid companies also maintained a small share of SP's generation assets.

- Five generation companies: China Huaneng Group, China Datang Corporation, China Huadian Corporation, China Guodian Corporation, and China Power Investment Corporation. These companies were each initially given around 20 GW of capacity, with the aim of ensuring that each had less than 20% of market share in any one region.

- Four power service companies: China Power Engineering Consulting Group, China Hydropower Engineering Consulting Group, China Water Resources and Hydropower Construction Group, and China Gezhouba Group. These entities combined key ancillary services that had been previously integrated into SP.

Separating assets: a fundamental step to reform

The separation of generation from the grid is not yet complete. Both main grid companies (SG and CSG) retain ownership of generation assets. In the case of CSG, this is only a small amount but SG still has over 30 GW, mostly hydro and coal-fired stations. Current plans are to divest some of the coal-fired plant.

Although care was taken in unbundling the State Power Company, various market share factors persist that are important to competition prospects. None of the five new generating companies were allocated more than a 20% share of any one of the regional power markets. However, their geographic roots in certain parts of the country are still evident. For example, Datang remains strong in the north, near the coal supplies; Huaneng is strong along the east coast; and Huadian is well-represented in Shandong province. It is also the fact that by developing majority shareholdings in consortia with other power investors, the five companies now control an amount of capacity – ranging between 30 and 38 GW – that is considerably higher than their original 20 GW allocation.

Thermal plants comprise 70-80% of the total capacity of all five companies, with hydro making up most of the balance. Datang has the lowest proportion of hydro whilst China Power Investment, at 30%, has the highest. China Power Investment is the only company with significant nuclear capacity, whilst Guodian is an important player in wind power.

Figure 3 Restructuring separated most generation assets from T&D assets

Generation assets

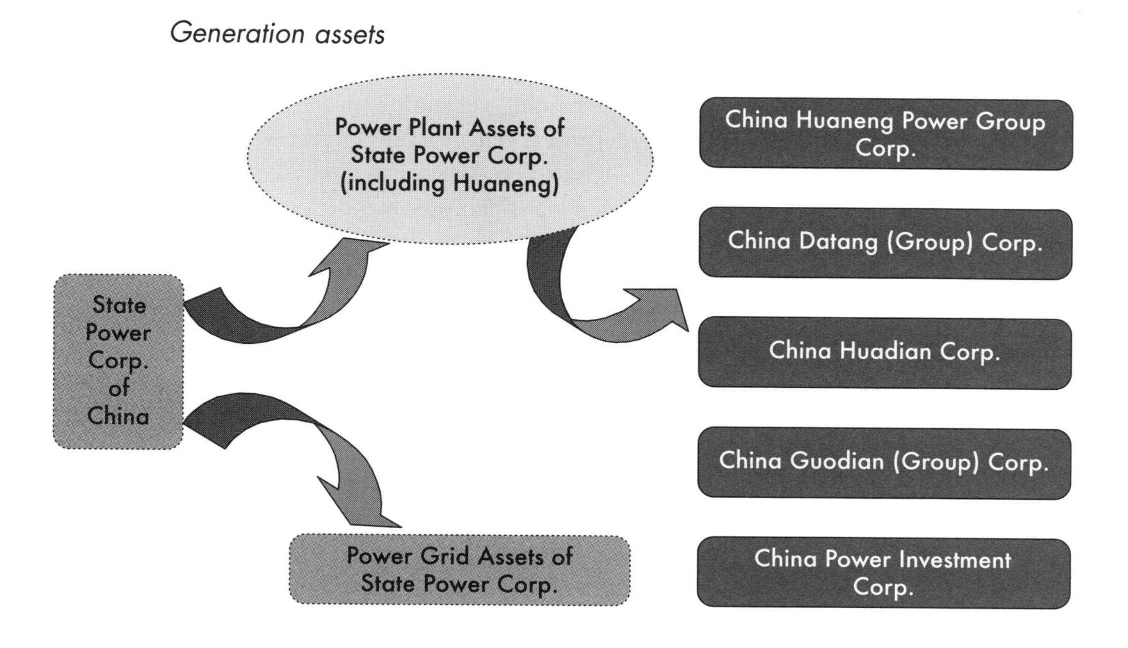

Transmission and distribution assets

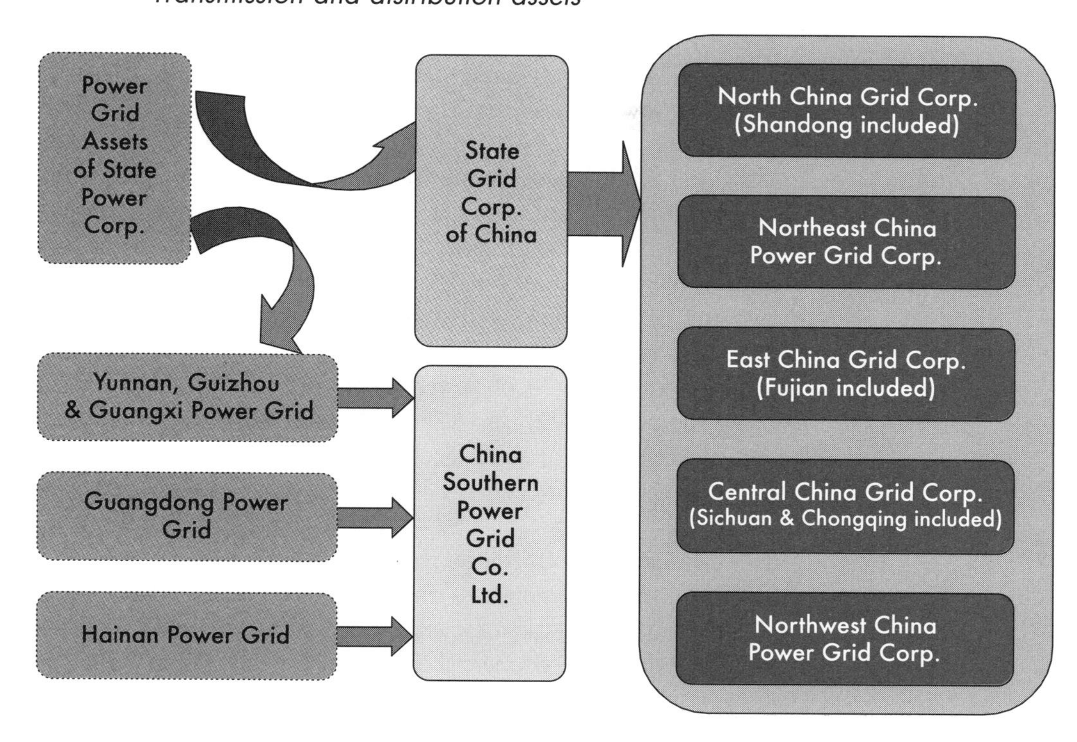

Central and local government retain significant ownership of the power sector's assets

The five large generation companies that were carved out of the State Power Corporation are, for the most part, state owned[4]. They also have majority shareholdings in consortia with other power investors, which extends their reach beyond the assets/capacity that they own directly.

Ownership of the remaining generating capacity is widely spread among industrial and financial enterprises, but remains largely with the state in various forms. Individual plants tend to be owned by consortia comprising various combinations of players, with or without the involvement of one of the five state-owned generating companies. Some of these players are at national level, such as the Three Gorges Dam, the Shenhua Group, and the State Investment and Development Company. But most are owned at local rather than national levels.

Partial privatisation through public offerings on domestic or overseas markets has brought in private assets. Some 40 joint stock companies have been partially floated on one or more stock exchanges. In about 40% of the companies the non-tradable block amounts to more than 10% of the total share capital. The size of the non-tradable shareholding ranges up to 35% and occasionally as much as 50%. In most of these listed companies the original owner retains a relatively low holding (15% to 30%) and the other non tradable shares are held by other legal entities. These other owners are nearly all SOEs at the provincial, city and sometimes national level. They include banks, asset management companies, investment, real estate and construction companies, as well as enterprises in the energy and chemicals sectors. Nevertheless, domestic private investors are playing a growing part in generation. Given their limited capital resources, this typically focuses on small projects such as small hydro, and often takes the form of joint ventures with local, government-owned corporations.

Diversification away from central government control and financing

Until 1985, China's power sector was under the direct control of central government, which owned all the assets. In that year, China liberalised investment by introducing a policy to encourage local government and companies to invest in new generation capacity, thereby gaining the right to control and benefit financially from this new capacity. The new policy led to a surge of investment at the local level, which lasted through to the mid-1990s. Funds from local government and other local sources accounted for just 14% of power sector investment in 1987, but averaged 40% during the period 1991-95. After 1995, the contribution of local funds to fixed asset investment in the sector declined in both absolute and relative terms, in part because of the growth of energy investment companies.

In the early days of reform, the source of central government funding gradually switched from the central government to the state-owned development banks. Then, from the early 1990s, the domestic commercial banks started to invest in the sector as well. By 2002, these domestic commercial banks were providing some 20% of total funds, with a corresponding decline in funding from development banks. The 1990s were also marked by three additional trends: the progressive involvement of a wider range of enterprises in power generation; the creation of power financing companies such as

4. Two of the five have had IPOs.

Huaneng and China Power International; and the increasing use of domestic and foreign stock markets. Combined with the corporatisation of the sector, this allowed power companies to provide a significant proportion of funding themselves. However, their contribution declined dramatically after the 1997 Asian economic crisis.

The role of foreign investment in generation

China also began to encourage foreign investment – but only up to a point and only for generation, where it accounts for less than 10% of the total. Initially, nearly all foreign investment took the form of funds provided by international financial organisations such as the World Bank and the Asian Development Bank, as well as national governments. From the late 1970s to the early 1990s more than USD 10 billion of foreign funds were used this way, primarily in the construction of power plants. But in the 1990s, against the background of demand projections that suggested the need for plant construction at the rate of 15 GW per year, the government decided to promote foreign direct investment (FDI). It set a target of 50% for the FDI contribution to projected needs, and planned to raise further foreign investment through stock markets. The 1990s saw a number of general, non-power-specific, measures to attract FDI. Within the power sector, the new approach to wholesale power tariffs, which distinguished between new capacity and existing plants, provided a further incentive. Approval processes were, however, a major bottleneck. Partly to counter this, build-operate-transfer (BOT) procedures were put in place to provide a system for managing investment in infrastructure projects in a consistent, efficient and transparent way, and to avoid the drawbacks of other means for bringing in foreign investment.

Foreign investors have typically taken one of three approaches to structuring their projects: co-operative (or contractual) joint ventures; equity joint ventures; or wholly foreign-owned enterprises. The co-operative joint venture provides the greatest flexibility for both Chinese and foreign parties, and has been the preferred structure. These three types of co-operative arrangements normally involve local power companies that are looking to boost their funds for new capacity. There is no formal tendering, and the "winner" emerges from prolonged opaque negotiations. The lack of control under these options has made them unattractive for potential foreign investors in China's power sector. The BOT scheme provided for an alternative route, by creating an open, transparent auction for the right to invest. The concession holder who emerges from a BOT auction has the right to construct a plant and run it for a specified number of years whilst receiving revenue for selling power under contract. When the concession expires, plant ownership goes to the state. The first BOT concession was awarded to Électricité de France (EDF).

Between 1995 and 1998, FDI successful helped to fund 21 major projects involving a total of USD 8.5 billion in foreign funds. Further projects are under discussion and an increasing range of foreign players are involved, though Hong Kong and other Asian investors continue to dominate.

Generation technologies, capacity and production

China's installed generation capacity reached 440 GW in 2004, having expanded at a rate of some 8% per year for more than 20 years (Table 1). During the 1980s, the proportion of hydropower capacity declined from 30% to 25%, but has remained relatively steady since then (Figure 4). In 2004, the sector saw a construction boom with an increase of 12% in total generation capacity, which compares to 5-6% in the years 1999-2002, and 9% in 2003. These official figures show a large increase in hydroelectric capacity during 2004, when several big projects came online. However, over the 15-year period from 1990-2005, fossil fuel (mainly coal) share remained significant and stable amongst primary movers (Figure 4).

Table 1 Installed capacity in China: 2003 and 2004

	2003	2003	2004	2004	2004/2003
	GW	%	GW	%	Increase %
Total	393		440		12.0
Hydro	94	23.9	108	24.5	14.9
Fossil	292	74.4	325	73.9	11.3
Nuclear	6	1.6	7	1.6	12.9
Other	< 1	0.2	n.a.	n.a.	n.a.

Source: Ye et al. (eds.) (2005).

Figure 4 Shares of installed capacity by prime mover, 1990-2005

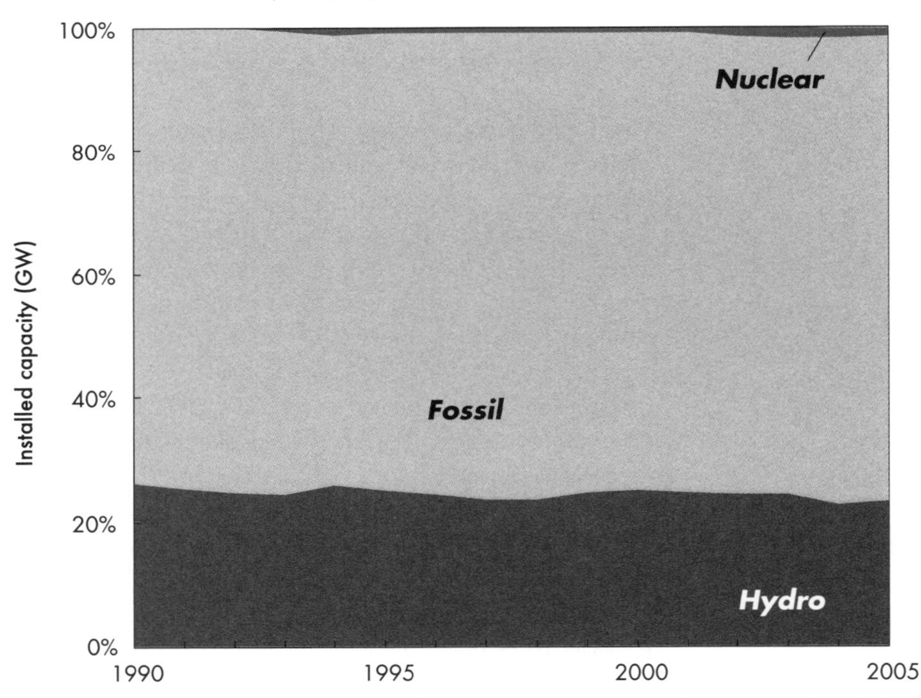

NB: Fossil fuel capacity is more than 90% coal-fired. Due to uncertainty regarding capacity of small diesel generators, an exact percentage is unavailable.
Source: Editorial Board of the China Electricity Yearbook (2004); National Bureau of Statistics (2006).

Raising the capacity of generating units used in new plants has been a key component of the national strategy for power generation. By 2003, 21 hydropower plants with capacities of 1 GW or more were in operation, including the Three Gorges Dam and the Ertan Dam. However, the proportion of hydro capacity in plants of 40 MW or greater is increasing only slowly and remains below 60% (Table 2).

Table 2

Proportion of hydropower plants with installed capacities of 40 MW and above

Year	1999	2000	2001	2002	2003
Number of units	311	337	355	361	388
Capacity (MW)	42.1	46.2	48.5	49.4	55.7
Shares in total capacity (%)	57.7	58.0	58.4	57.4	58.7

Source: Ye et al. *(eds.) (2004).*

The trend towards increasing scale is more evident among fossil-fired plants. By the end of 2003, 83 plants with a capacity of 1 000 MW or more were in operation, seven of which were commissioned during the same year. The proportion of units with capacities of 300 MW or more increased from 36% in 1999 to 43% in 2004 (Table 3). Most new approved projects have a minimum capacity of 300 MW, and the government encourages the construction of plants with capacity of 600 MW or more. Despite this trend, some 4 000 units, representing 20% of installed capacity in 2003, had capacities of 50 MW or less.

Table 3

Proportion of fossil plants with capacities of 300 MW and above

Year	1999	2000	2001	2002	2003	2004
Number of units	238	262	295	313	339	339
Installed capacity (MW)	80.4	90.9	103.2	110.8	119.9	143.0
Shares in total fossil power (%)	36.0	38.3	40.8	43.2	41.4	43.4

NB: In 2004, about one-quarter of new thermal capacity additions were less than 25 MW, presumably mainly small diesel generators
Source: Ye et al. *(eds.) (2004 and 2005).*

Fuel technologies: coal-powered thermal dominates

Currently, 82% of China's total power is generated from fossil fuel combustion with hydroelectricity supplying a further 16%. China has six nuclear power plants with a total capacity of 7 000 MW, amounting to 1.6% of total generating capacity. More plants are coming on line quickly. Non-hydro renewables, mainly wind, account for less than 1%.

Among the power from fossil units, coal supplies over 90%. Up until the end of 2003, use of natural gas in power generation was limited, accounting for only 5 000 MW (1.7%) of total fossil capacity. Gas-fired capacity has since more than doubled, but much of it is idle due to lack of fuel. Approximately 4-5% of total thermal fossil capacity is fuelled by oil. Diesel-fuelled plants are especially common in coastal provinces such as Guangdong, which has a long history of power shortages.

The structure of generating capacity varies significantly from region to region (Table 4). At the extremes are central and south China where hydropower is plentiful and north and northeast China where coal is abundant and where fossil power is highest. Nuclear power is restricted to the richer, energy-deficient southern and eastern areas.

Table 4 Prime movers shares for power generation, by region, 2003

	Hydro	Fossil	Nuclear	Other	Total
Installed capacity (GW)					
North China grid	3.0	81.8	0.0	0.1	84.9
Northeast China grid	5.9	35.3	0.0	0.2	41.3
Northwest China grid	9.1	20.7	0.0	0.1	29.9
East China grid	13.6	65.2	2.4	0.1	81.3
Central China grid	36.1	47.4	0.0	0.0	83.5
Southern China grid	26.0	41.6	3.8	0.1	71.5
Total	94.0	292.1	6.2	0.6	392.8
Shares of capacity					
North China grid	3.5	96.3	0.0	0.1	100
Northeast China grid	14.3	85.5	0.0	0.5	100
Northwest China grid	30.4	69.2	0.0	0.3	100
East China grid	16.7	80.2	3.0	0.1	100
Central China grid	43.2	56.8	0.0	0.0	100
Southern China grid	36.4	58.2	5.3	0.1	100
Total	23.9	74.4	1.6	0.2	100

Source: Ye et al. (eds.) (2004).

The period covered in the next Five-Year Plan projects that coal-fired plants will continue to play a major role (72%) in capacity expansions. Hydro would account for another 20% and nuclear for 1.2%, with the balance being accounted for by natural gas and oil (most of which would be natural gas). The government is seeking to promote natural gas, though this is from a very low base and requires major efforts to expand pipelines and other infrastructure. Total gas-powered generating capacity may rise to 20 GW or more by 2020. Power shortages may provide new incentive for nuclear power, but its future remains in doubt: high costs and financing difficulties have prevented the technology from meeting its original goals. Current plans anticipate expanding nuclear capacity from 7 GW to 12 GW in 2010, and to 40 GW by 2020. China is investing heavily in large hydroelectric projects, mainly in the central, southern and north-western areas of the country. The Three Gorges Dam had about half the planned units in operation by mid-2005, with a final completion date of 2008. Another 40 GW of capacity will come from other large hydroelectric projects in the Southwest, which are under construction or in advanced planning.

The grid and system dispatch

Despite recent investments, China's transmission grid system remains weak. The provincial roots of China's original power industry led to a system comprising a large number of separate, high-voltage transmission grids. The number of separate grids has decreased over time – from 18 in the early 1980s to 10 in 1997 (mostly through interconnection), and to six main regional grids by 2004 (see Figure 1 for locations of provinces):

- The North China grid covers Beijing and Tianjin municipalities, Hebei, Shanxi, Shandong provinces, and the west part of Inner Mongolia.
- The Northeast China grid covers Liaoning, Jilin, Heilongjiang provinces and the east part of Inner Mongolia.
- The East China grid covers Shanghai municipality, Jiangsu, Zhejiang, Anhui and Fujian provinces.
- The Central China grid covers Henan, Hubei, Hunan, Jiangxi and Sichuan provinces, and Chongqing municipality.
- The Northwest China grid covers Shaanxi, Ningxia, Gansu, Qinghai and Xinjiang provinces.
- The South China grid covers Guangdong, Guangxi, Guizhou, Yunnan and Hainan provinces. However, Hainan province is not yet connected with the others.

Interconnections have developed rapidly since 2000; five of the six main grids are now interconnected (Figure 5). In addition, a 500 kV AC transmission backbone (4 600 km in length) links Northeast China with North China, and North China with Central China, tying together almost 200 GW of generation capacity. Central China is now also connected with the East China and South China grids through 500 kV DC lines. In July 2005, the Northwest China grid, which has a lower maximum voltage of 330 kV, was linked with the North China Grid. However, some provinces, such as Fujian, are not well interconnected to other provinces and two regional grids (Xinjiang and Tibet) remain completely isolated, as does the island province of Hainan.

These developments reflect major grid investments over the last ten years. Before 1995, grid development substantially lagged behind growth of generation capacity. Grid investments reached 44% of total fixed asset investment in the power sector during the 9th Five-Year Plan (1995-2000), compared to 20% between 1985 and 1995. Total transmission investment during 1995-2000 (220 kV and above) was reported at USD 17.2 billion (RMB 143 billion *yuan*)[5], and urban and rural distribution investment at USD 24 billion (200 billion yuan). The investment in the grid system in 2001 and 2002 was USD 12.5 billion (103 billion yuan) and USD 13.8 (114 billion yuan), respectively. The two state grid companies continue to invest heavily in grid construction, allocating respectively USD 3.4 billion and USD 10 billion (28 billion yuan and 83 billion yuan) in 2004. (Foreign involvement in grid investment is explicitly forbidden.)

5. The Chinese currency is referred to as *Renminbi (RMB)* and is denominated in *yuan*. As of this writing, one Euro is worth 10.05 yuan and one USD (dollar) is worth 8.01 yuan. Figures in USD reflect the exchange rates that were in effect during the years referenced. In this report, we use the term yuan and hereafter drop the initials RMB.

The main grid system does not yet cover the entire population. How far it should do so is an issue that needs to be considered, particularly in light of the contribution that can be made by off-grid renewables and distributed generation.

Investments have also been made in the distribution grid. Between 1998 and 2002, China carried out an urban and rural distribution network construction and reconstruction project, investing approximately USD 47 billion (390 billion yuan) in 269 urban distribution networks and 2000 county distribution networks. As a result, the distribution network is greatly improved through increased quality and reliability of electrical power supplies, especially in rural areas.

Despite such investments, the map in Figure 5 shows that much additional construction remains to be done in the development of a strong, interconnected grid.

Further development aims to correct important weaknesses

China continues to face a significant challenge in that many of the country's rich coal and hydroelectric resources are located far from populous markets and centres of economic activity. Considerably more investment is needed to strengthen the grid and develop interconnections that will allow power, which is largely generated in the west, to reach demand centres, largely located in the east.

At present, grid bottleneck are the main cause of supply shortages. The bottlenecks are, in turn, due to the limited transmission capacity between regional grids or with weaknesses in the transmission of bulk power at local levels. These bottlenecks compromise the availability of power on a much broader scale. For example, Northeast China experiences congestion between Heilongjiang province, which has a surplus of generating capacity, and Liaoning province, which faces power shortages. In the East China region, capacity is not sufficient to support transmission from the northern part of Jiangsu province, which has a power surplus, to the southern part of the province.

Two additional issues are worth noting. Despite the establishment of a synchronised inter-regional grid, interconnections are still weak. This influences the stable operation of the regional grids, and sometimes reduces the stable limit of bulk power transmission within a region. In addition, there is still a significant technology gap between China's grids and those of developed countries. More advanced technology should be adopted to help raise the level of stable operational grid capacity.

Chinese authorities recognise all of these weaknesses and have, therefore, included further investment in the grid as a key component of future plans for the power sector (Box 3).

Figure 5 Major interconnections between China's regional grids, 2004

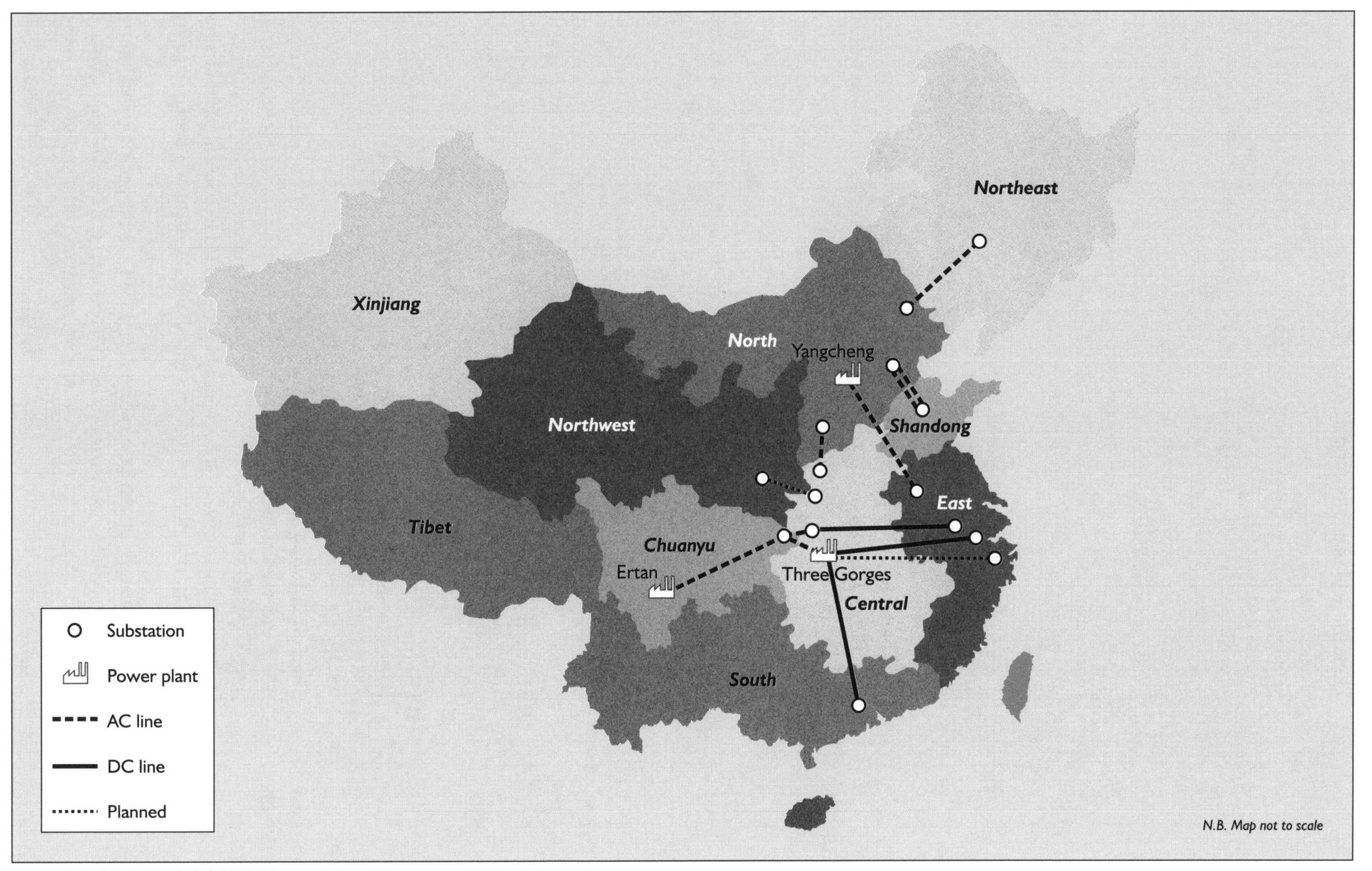

Source: Based on Ye et al. (eds.) (2005).

Box 3 Plans for grid strengthening and development

China's central objectives for increasing generation capacity are closely linked to regions and resources. Plans for generation focus on developing coal resources and adjacent power generation facilities in the west and hydropower in the southwest. This will be complemented by a plan to enable this power to be carried to the demand centres in the east, as set out in the government's 2000 strategy of West-to-East electricity transmission, and confirmed in the State Council's 2002 power reform document. The latter sets out two key initiatives related to its broader goal of construction of a unified national grid: further development of the West-to-East transmission system and strengthening inter-regional interconnections. Specifically, the plans envisage:

- **Further developing and enlarging west-to-east transmission capacity,** including plans for three channels:
  - A Northern channel, including transmitting thermal and hydro-electricity from the Northwest to the North region, and the transmission of electricity of thermal power from Shanxi province and western area of Inner Mongolia to Beijing, Tianjin and Hebei provinces within the North China grid.
  - A Central channel, including the development of Three Gorges and other large power stations on Jinsha River and in Sichuan provinces to transmit electricity to the East China grid, which is facing severe energy shortages due to rapid economy growth.
  - A Southern channel, which will mainly develop the hydropower plants and coal pit power plants in Yunnan and Guizhou in order to transmit electricity to Guangdong province.
- **Strengthening interconnections between regional grids** by building north-south interconnections between the developing west-to-east transmission lines.
- **Strengthening and optimising the development of bulk power transmission** within each region, including strengthening the regional backbone transmission lines and improving provincial transmission systems.
- **Continuing to improve urban and rural distribution networks,** thereby increasing the total supply capability of distribution networks.

Transmission and distribution ownership

The 2002 reforms established the two state-owned grid companies, SG and CSG, which are responsible for managing the grid assets in their territories, and for inter-regional system operation. For operational purposes, each company is divided into smaller management areas, which have delegated responsibility for local network development, maintenance, system control and dispatch, and system security.

SG is the larger of the two companies, owning and operating the interregional transmission lines, and five of the six regional grids[6]. Although some of the distribution assets are owned by local government and other entities, SG owns about

6. SG also manages the Tibet network under the authority of the government of the Tibet Autonomous Region.

75% of the transmission and distribution lines in its service area and about 88% of the transformers. CSG operates the southern regional grid, which it jointly owns with the provincial governments of Guangdong and Hainan.

SG directly owns the interconnections between the regional grids. It is also responsible for inter-regional trading and the operation and development of inter-regional grids. The regional grid companies, including CSG, are responsible for the development and operation of the regional grids, regional system dispatch, and the development and operation of the regional power markets.

A number of provincial/municipal companies are contained within the framework of each regional grid company and CSG. They typically own and operate the transmission network within the boundary of the province. They are the sole buyers of power from the generation companies, and are responsible for resale to consumers and distribution companies within their franchised areas.

System dispatch

System dispatch is managed at three levels: inter-regional, regional and provincial. The dispatching centre within SG is in charge of all the interregional transmission lines and facilities. Regional dispatching centres manage transmission dispatching within each region. Provincial dispatching centres oversee scheduling to implement yearly contracts and to conduct real-time balancing to control provincial power systems.

Generator dispatch in areas of the country that do not have competitive power markets (*i.e.* most of China) is done on the basis of an average tariff level for each plant, which is approved annually by NDRC. Generators are paid a single energy-based price (*i.e.* per kWh) for their output, which is intended to cover the annual capital and operating expenses. Dispatching of plants depends on a combination of this price and a preset allocation of operating hours, which varies according to the plant. Specifically, the grid company schedules a month-long load curve, according to historic patterns of usage and load forecasts. Each plant is dispatched according to its type, assumed operating hours, and forecast load curve.

Wholesale and retail sales and trade

For the most part, China operates under a single buyer model. Power sales are growing between regions but are still carried out through tightly planned transactions. Generation is sold to the grid companies under long-term contracts that are set and approved by the government. In turn, the grid companies sell the power to end users, again under government-approved retail tariffs.

Construction of new transmission lines has enhanced interconnection between grids, leading to increased power sales between regions[7]. But again, these are tightly planned and managed transactions. For example, all electricity from the Three Gorges dam is sold according to a plan and a price set by NDRC. The sale price between Northeast China and North China is negotiated by the two regional grid companies, and a transmission plan is made at the beginning of each year, when the electricity price is negotiated. The network service fees are collected by SG based on the price

7. For example, from Northeast China to North China, through a 1 200 MW AC line; from Central China to East China, mainly from the Three Gorges and Gezhouba dams through a 4 200 MW DC line; from Central China to South China, from the Three Gorges dam to Guangdong province; from the Yangcheng power station in North China to the East China grid.

regulated by NDRC. Similarly, transactions between Yangcheng power station and Jiangsu province (located in the East China grid area) are jointly regulated by NDRC and Jiangsu provincial planning commission.

The Yangcheng power station presents an interesting case. It is situated in Shanxi province, China's main coal production base, but is committed to selling its power to Jiangsu province to which it is connected by a dedicated AC transmission line. Thus, it is treated as a power station within Jiangsu province and pays network fees to SG at a regulated rate.

The Central China grid and the East China grid engage in a number of short-term transactions. The price is negotiated by the two grid companies, but the approval of SG is needed for access to the required transmission capacity.

Organised power markets

In the late 1990s, an over-supplied power market provided an excellent context in which China could conduct early experiments in competition and organised, wholesale power trading. The experiments broadly followed the British mandatory power pool model (pre-NETA), requiring qualified generators to sell power through a single buyer (the regional grid company) in four regions: Shanghai, Shandong province, Zhejiang province and the Northeast (Jilin, Heilongjiang and Liaoning provinces). Typically, each province took a small fraction of market demand and selected a certain number of generators to participate in the competition to meet this demand. Generally, the 12 largest IPPs in each province were required to participate by bidding roughly 10% of their normal contractual delivery to the pool. Tariffs were capped and no actual financial settlements were undertaken. In reality, planned allocated dispatch continued to meet the bulk of the demand.

The experiments were short-lived. The oversupply of power quickly dissolved and all available power was returned to the allocated system. Moreover, they fell short of reflecting a real market. There was little independence in the process, and in particular, no independent regulatory oversight. In addition, there were assertions that the State Power Corporation made many decisions in favour of its own generators.

In 2002, under the State Council reform plan, China began to roll out regional power markets.

One year later, SERC issued the *Guidelines for Establishing Regional Power Markets*, a document that described the objectives, the main models and the main trading approaches for the proposed markets. In 2004, the first two markets were launched, on a pilot basis, in Northeast and East China; a third, in Southern China, was launched in 2005. The government's main objectives are three-fold: to establish a unified, open, competitive, and orderly power market; to break down provincial protectionism; and to stimulate investment. It also notes that the power markets must move forward in a prudent manner, and should adapt to the special situations in their own regions.

The aim is to extend this process to the three other regions. Each market will be developed in several stages, starting with a simulation in which the bidding system is put into operation but there is no actual settlement. The first stages will be quite restricted in terms of market participation, contractual mechanisms, and the absence of financial markets.

Trading through the markets will not be mandatory. However, once they have evolved, these markets will include key features such as supply-side bidding by a majority of generation companies in the region, demand-side bidding from qualified large end consumers (*e.g.* independent companies as well as distribution companies), and provision for bilateral transactions. The trading mechanisms will include yearly contracts, monthly bidding contracts, day-ahead bidding and real-time balancing. Eventually, there are plans to extend choice to all retail customers (at least for the Northeast China market).

Market participation will be broad, including coal, oil, natural gas, nuclear and hydropower stations. Wind, geothermal and other new and renewable forms of energy will be subject to separate rules. Foreign invested power plants approved and constructed before 1994, which have signed power purchase agreements or which have received other government undertakings, will be obliged to renegotiate.

Regional differences

The reason China has chosen a regional roll out is relatively straightforward. The differences of economic development – not only across the country but within regions – effectively rules out the possibility of applying an identical approach to each market. Moreover, it is difficult to implement a unified pricing system because the poorer provinces may not be able to afford a higher price, and some regions have special characteristics, such as hydro. Because of their distinct natures, the Central and South China markets will require a different approach from that used for the first two (Northeast and East China). The Central China market needs to deal with the competition of hydropower, which accounts for one-third of total generating capacity. The South China market needs to focus on the huge economic gap between Guangdong province and the other three poorer provinces.

Northeast China power market

The Northeast China power market – which is headquartered in Shenyang and includes the provinces of Liaoning, Jilin, Heilongjiang, and parts of Inner Mongolia – was launched in January 2004. Some 20 generators were originally selected to participate in the market, with 15% of their normally allocated volumes to be bid into the market, using a one-part price model. In June 2004, as a means of encouraging investment, this was changed to a two-part price model with all electricity bid into the market. Electricity is currently bid for month-ahead scheduling, but could later be expanded to include daily or real-time trading.

There were two factors behind the choice of Northeast China as a test market: *(i)* Retail tariffs and the level of economic development are similar across the three provinces in the region; and *(ii)* It is the only one that has experienced a power surplus in the last two years. This market will be implemented in three phases, with specific activities identified in each phase:

- **Phase 1:** Establish a trading system and market supervision system, and form a unified trading centre; open part of the generation market for competitive bidding;

and establish a market with two-part pricing (regulated capacity price and competition-based energy price).

- **Phase 2:** Establish an ancillary services market; extend competition coverage; begin trial bilateral transactions between generators and independent distributors or large consumers; and form a unified regional power market across Northeast China.

- **Phase 3:** Introduce retail competition; establish an electricity money market; and establish a generation capacity market.

East China power market

The East China power market was launched in May 2004. Bidding into the market is compulsory for qualified generators, which covers coal fired plants with capacities of 100 MW or greater. It will also be implemented in three phases, with specific activities planned in each phase:

- **Phase 1:** Establish a unified regional platform for electricity trading; and initiate a gradual move to allow large consumers to purchase electricity directly from generators.

- **Phase 2:** Promote bilateral trading between generators and large consumers or independent distribution and retail companies; open ancillary services and transmission rights trading markets; improve management and supervision system for market pricing; and form a market with unified operation in the region.

- **Phase 3:** Introduce competition in the retail market, with all qualified generators and consumers participating; develop financial trading; and form a unified, open, competitive and orderly regional market under government regulation.

THE GOVERNANCE AND INSTITUTIONAL FRAMEWORK

China's governance and institutional framework for the power sector is in the midst of a long, drawn-out process of transition. The transition is linked to wider reforms undertaken in the 1980s to promote growth and economic development, although the rate of change in the power sector has been much slower than in most other industries. Considerable reform has taken place over the last 20 years, including important changes to the governance, structure and ownership of the power sector (Box 4)[8]. These reforms have laid a potentially effective groundwork for further development, and further reform proposals have already been drawn up.

By 2005, China had moved from a single vertically integrated utility to two grid companies (a large one covering most of the country, and a small one in the south), and a diverse set of generation companies (including five large companies that were spun off the original incumbent). Competitive power markets have been launched in two regions, and more are planned. In addition, a first step has been taken towards independent regulation with the establishment of a regulator, the State Electricity

8. See Box 2 on market structure and ownership for a more detailed review of industry unbundling.

Regulatory Commission. However, ownership of the power sector is still largely with the state: the grid is owned by central government, and generation companies are often linked with local state interests, either directly or indirectly.

Box 4

Need for reform: progress to date

In 1993, China invited The World Bank to offer advice on power sector reform. The World Bank (1994) recommended that China take steps to[9]:

- Separate the industry from government, and to corporatise and commercialise the industry.
- Develop an economically rational system of power pricing that provides appropriate signals to investors in generation, transmission and distribution, as well as to consumers.
- Embark on the staged development of a power market by separating generation from transmission and transmission from distribution, and to create a number of generators selling to a single purchasing agency. This would be followed by the creation of power markets.
- Create a legal and regulatory framework for implementing these reforms and for regulating the evolving power market. This would include the establishment of a regulatory commission at national and provincial levels.

The reform path proposed by the World Bank has been broadly followed as regards restructuring. However, pricing reforms, a key element, have been delayed, and the other reforms are as yet incomplete. Governance and institutional capacities remain weak, despite the establishment of a regulator.

Current policy: the 2002 State Council Document

In April 2002, China's State Council (the country's highest executive authority) issued a document that set out major reform plans (General Office of the China Secretary of the State Council, 2002). The stated goals were: to encourage the long-term development of the power sector; to ensure a safe, efficient and reliable power supply; and to protect the environment. To this end, the plan called for:

- The break-up of the State Power Corporation into five new generating companies and two national grid companies.

9. Subsequent reports have examined specific dimensions of the proposed reform process in more detail, including the development of competitive power markets, power sales between markets, how to structure and empower the regulatory commission, and how to manage complications arising during the transition process. The US Energy Foundation has been another major adviser to the Chinese government on reforms.

- The establishment of a regulator (the State Electricity Regulatory Commission) under the State Council to regulate the developing markets.
- The establishment of competitive power markets for generation across the regions, overseen by SERC. The original goal was that major generators should participate in competition by the end of 2005.
- Development of a new pricing system.
- Incentives for clean energy development and new power plant emission standards.

The first part of these reforms has been implemented, and two regional power markets have recently been launched in east and northeast China. A new pricing system has not yet been implemented.

Institutional developments, 1998 to present

Prior to 1998, the basic institutional structure of China's power sector comprised a set of ministries or very large state-owned enterprises (SOEs) that owned the energy sector's assets. These entities were responsible for all aspects of the energy sector and its development, including asset management, investment, planning, and price regulation. They reported to the State Planning Commission (SPC) and the State Economic and Trade Commission (SETC) which, respectively, held responsibility for strategic and operational matters. Most importantly, the SPC was required to approve all major investments and all energy prices.

While broad reforms began much earlier, 1998 saw the start of important efforts to separate corporate responsibilities from government oversight of the energy sector, including the power sector. The key government functions relating to the energy sector were allocated to the renamed State Development Planning Commission (SDPC), the SETC and the newly created Ministry of Land. At the same time, the Ministry of Electric Power was abolished and its assets transferred to the new State Power Corporation.

Four years later, the establishment of SERC as a regulatory body (by an executive order of the State Council[10]) marked another important step. SERC's mandate was to oversee the more competitive company structure that was developing and to promote further power sector reform. It reported directly to the State Council. The State Council, in its 2002 policy document, set out the following responsibilities for SERC: establish and oversee market rules, including competitive bidding rules and protecting fair competition; make tariff modification proposals; monitor production quality standards; issue and monitor licences; settle disputes; and oversee implementation of universal service reform. At the same time, the government established a new State-owned Asset Supervision and Administration Commission (SASAC), which held responsibility for management of the state-owned power companies.

10. The State Council oversees all State agencies and sets China's broad policy directions and priorities.

As a result of these reforms, the power industry is now governed by three key institutions:

- NDRC, which itself has three relevant bureaus:
  - Energy Bureau. This bureau is responsible for power sector policy and strategy, and approves projects.
  - Pricing Department. This department sets prices for the power sector.
  - Environment and Resources Comprehensive Utilisation Department. This department is concerned with energy efficiency.

- SERC, which is, in principle, responsible for regulating the sector (apart from setting prices and project approval) and developing reform plans. It has branches in the regions.

- SASAC, which is the shareholder for power sector SOEs, and is responsible for overseeing their management and boosting performance.

The State Council issued a notice in February 2005 that strengthened the powers of SERC. The notice explicitly mentions the supervision and administration of non-discrimination and of market rules, and lists the measures SERC may take to carry out its responsibilities, especially with respect to gathering information. It also specifies the legal liabilities of SERC and of individual SERC staff with respect to inappropriate or illegal behaviour. There is currently discussion regarding the possibility of setting up a new energy ministry, but this has not yet taken shape.

Pricing regulation

China's current pricing regulation needs to be understood in the broader context of its single buyer model for power transactions. Generators sell power to the grid companies at regulated prices, and the grid companies sell the power on to end users, whose prices are also regulated. There is no separate pricing for the grid; it is embedded in the other prices. The approach also applies in the recently established regional power markets, in which end users are not yet allowed to transact directly with the generators. In short, the value chain is currently linked through the grid companies acting as single buyers.

The evolution of wholesale pricing regulation

Before 1986, nearly all power plants in China were constructed with funds from the central government. The wholesale price of power was set according to approved Catalogues published by the central government, which took into account operating costs but not capital costs.

Beginning in the late 1980s, the government started to allow investment from other sources. New plants were allowed to charge higher prices to recover costs and to provide a fixed return on profit[11]. The aim was to encourage investment from non-government sources. Initially, the higher prices applied to plants constructed between 1986 and 1992 that did not use central government funds. After 1992, all newly constructed plants could charge higher prices, regardless of the type of

11. In the late 1980s, the government also began imposing fees to collect funds for developing selected generation and grid projects. For instance, a small per kWh charge was levied on all electricity sales nationally, from both old and new power plants, to finance the Three Gorges Dam Project.

investor. The "new price" was set separately for each plant's debt-servicing period and for the period after debt repayment.

The "new price for new power" concept evolved into a system under which wholesale prices were set by the government (usually at provincial level, by the Provincial Pricing Bureau) with final approval from the central government's State Pricing Bureau. The price would be based on the age, efficiency, fuel, location and type of power generated (peak or off-peak). Substantial differentials persisted. For example in 2001, the average price paid to generating plant constructed before 1985 was USD 0.029/kWh (0.24 yuan/kWh). For new plant with prices approved in 1997, it was USD 0.049/kWh (0.41 yuan/kWh). This "new price for new power" scheme was successful in encouraging investment, and during the 1990s the numbers of parties investing in power generation multiplied, as did the numbers of plants. The system, however, provided no incentive for investors to reduce their costs or to seek more favourable financing terms.

Today, both old and new plants are paid on an output basis, *i.e.* per kWh produced. Their annual capital and energy costs are recovered in these energy based prices. The capital cost component of the price is calculated on the assumption that the generator will operate for a specified number of hours per year, which is then allocated on a monthly basis. For the purpose of fairness and equity, each generator of the same type operates roughly the same hours per year, regardless of operating costs or fuel efficiency.

In 1998, the government introduced a new policy known as the "operating period tariff". This policy set the tariff on the expected lifetime of the plant, rather than on the debt repayment period. The lifetimes were set at 20 years for fossil fuel plants and 30 years for hydro. The assumed return on equity was set at 2-3% above the long-term bank lending rate, and the costs of each plant were benchmarked against plants of similar fuel, age and unit size. The objective is to control and lower the capital cost of new plants and to place the responsibility for negotiating suitable financing terms on the project sponsors. An additional component of the policy is to pay a "transitional tariff" to plants in their initial years of operation – before setting the "operating period tariff". It is expected that the "transitional tariff" will be lower than the "operating period tariff".

In 2004, another new pricing policy was adopted, in part to encourage improvements in efficiency. The prices paid to new generators are set on the basis of technologies in each province. For example, within a province all new coal fired generation with flue gas desulphurisation (FGD) is paid the same price, and this price is different from the price paid to other technologies such as hydro. Each price is based on current estimated construction and operating costs of the various technologies, specific to the provinces in which they are located. This is, in effect, a kind of standard-offer pricing[12].

12. It is similar to some United States pricing for the promotion of renewables, except that in China's case prices may be adjusted periodically by the NDRC to reflect issues such as fuel prices.

Transmission and distribution pricing

At present, transition and distribution (T&D) pricing is not based on transmission costs. In addition, intra-utility transmission, which is the largest share of bulk transactions, is not a separately priced service charged to generators. Instead, T&D costs are bundled in the retail prices charged to end users. In essence, the residual between existing retail prices and generation prices covers transmission, distribution and the remaining (non-generation) functions of the power companies.

In the case of independent distribution companies that provide retail service, they purchase power (in addition to any they supply themselves) from the large provincial utilities in their area. The prices for this power are set by the NDRC and typically include a mark-up over the wholesale generation cost. However, this mark-up does not necessarily reflect the actual costs of delivery.

In regional generation markets, there are tariffs for transactions between regional grids and for transactions within regional grids. Inter-regional transmission tariffs use a combination of capacity and energy charges. Intra-regional transmission tariffs generally use only capacity charges. It is not clear, however, that these prices are – in either case – cost reflective.

End-user pricing

The Catalogue system for consumer tariffs started in the 1960s as a means of giving preferential treatment to heavy industry, chemical plants, agriculture and irrigation – in terms of both the allocation and price of power. It has evolved to cover eight main categories of consumer with three voltage classifications, making 24 basic categories. The Catalogue forms the basis of end user tariffs throughout China. Each of the categories is assigned a Catalogue price, which is used to calculate the final price.

A range of charges and fees, which have changed over time, are added to reach the final price. The main charges and fees today are for:

- Debt payment for urban and rural network construction – about USD 0.0024/kWh (0.02 yuan/kWh).
- Three Gorges Dam construction – from USD 0.0005 to 0.0018/kWh (0.004 to 0.015 yuan/kWh) for different consumers in different regions.
- Urban Utility surcharge – from USD 0.00024 to 0.0024/kWh (0.002 to 0.02 yuan/kWh) for different consumers in different regions.

Time-of-day tariffs are applicable to all categories of users – except residential and irrigation. However, the differentials, which are based on the Beijing Catalogue, are low. The maximum differential between peak price and standard price is 1.7, whereas a value of 4-6 would be more effective. The differentials for heavy and preferential industry are significantly lower than for other users. The low differential is clearly aimed at supporting such industries, which have traditionally received subsidised power.

Another pricing mechanism is differential voltage charges. The discount offered to Beijing customers that take power at a higher voltage is consistently less than 5% of the Catalogue tariff for the adjacent voltage band. This is only a small difference, with the effect that the high voltage users, *i.e.* users that have to build and pay for additional transformer facilities, subsidise the low voltage users.

It is not surprising that a country as vast and diverse as China would also have local differences in end-user pricing. Each province and major municipality amends the Catalogue Price to suit its own policy goals and economic development, and may add a different suite of additional fees to the nationally approved ones. This means that power prices can vary considerably across China. They are, for example, lower in central and western China than in the south and east.

Considerable pricing differences can even be found within Beijing itself. The Beijing Catalogue prices covering the period 1997-2004 show that prices are relatively low for domestic users, heavy industry, preferential industry, agriculture and irrigation, and relatively high for non-residential lighting and the commercial category. Prices were increased above inflation in the late 1990s, though this did not affect residential consumers. However, a 2002 round of price increases hit the residential sector hardest, thereby implying some unwinding of cross-subsidies to this category. The power shortages of 2003-04 triggered a further round of price increases. These were directed mainly at heavy and preferential industries, which experienced a 10-15% rise. The other categories received proportionally smaller increases, with residential and agricultural consumer prices remaining unchanged. Overall, Beijing residents remain highly protected.

At the same time, there remains a degree of discrimination against rural consumers. Each province has unified prices for the same type of consumer, regardless of location within the province. Current prices consist mainly of per kWh energy prices, differentiated mainly by customer groups, and a demand charge for large industrial consumers.

For many years, rural consumers were subject to a variety of legal and illegal additional fees, which had the impact of adding 100% or more to rural tariffs in comparison to urban tariffs. In 1998 the central government launched a drive to eliminate such discrimination. This has, reportedly, been successful. Further, in 2000 and 2001, the "new power fee" and "new connection" fee were banned.

Coal pricing for power

Coal has long been sold to power generators at prices significantly below market levels. These prices are agreed at the annual National Coal Procurement Conference run by the NDRC and its predecessors. The system brought a degree of predictability to the market and appeared to suit all the parties – at least until 2002. By this time, coal supplies were beginning to tighten as economic growth accelerated and the coal industry faced a capacity shortage following the widespread enforced closure of small-scale coal mines. The domestic market price for coal doubled between 2001-04. From 2002 onwards the annual meeting became progressively more acrimonious, as producers sought to capitalise on rising market prices and generators fought to protect their diminishing margins. As coal prices have risen, it has become more difficult for the parties to reach agreement, mainly because generators have had no means to pass through price rises to end users.

Although a "coal-for-power" formula is still applied, coal prices are in effect now largely determined by the market. They rise when supply is tight and fall at times of over supply. In December 2004, the NDRC announced a new scheme for linking wholesale power prices to coal prices. The link is defined by a formula that includes standard coal consumption and the calorific value of the coal. It provides for approximately 70% of a rise in the coal price to be passed through to the client. A change in coal price of 5% or more will trigger an immediate adjustment of wholesale prices. Lesser changes of the coal price will be addressed in six-monthly reviews. This should go some way to satisfy the needs of the coal mining enterprises and the generators, as it will partly remove current constraints on, and uncertainties in, their respective revenues. What has yet to be determined is how these price rises will be passed on to end users. Until this is resolved, the generators stand to lose revenue.

The rise in coal price should provide further encouragement for coal mine enterprises to produce more coal and to invest in new production capacity. However, continued coal transportation bottlenecks will mitigate these incentives and likely reduce the availability of coal.

In the longer term, planned retail price reforms include a corresponding mechanism to adjust retail prices. The aim is to reduce the risk of fuel price volatility to utilities, by shifting the risk to consumers.

Proposals for further reform

Much remains to be done in the area of reforming governance and institutional framework, as well as in the area of pricing. Current proposals centre on further pricing reforms. Other reform proposals are being debated but their content is not yet confirmed. For example, there are moves to establish an energy ministry or office, but its role and relationship with existing entities, such as the NDRC, is not yet clear. The roll out of the regional power markets is also the subject of ongoing debate and fine tuning.

Proposals for pricing reform

For the first time, recent proposals advocate separate tariffs for transmission and distribution, based on a "postage stamp" principle. Initially, three separate sets of tariffs will be created: for generation, for transmission and distribution, and for retail. The tariff for transmission and distribution will be set on the basis of cost recovery, reasonable profit and tax liability. Initially, the "postage stamp" approach will be used, under which all grid users in a given region pay the same tariff, regardless of the costs generated by their location. A separate "service" tariff, which includes the connection fee, will also be charged. Formulae are provided for the calculation of permitted profit and capital cost. Eventually, transmission and distribution tariffs will also be separated.

In addition, there will be a two-part wholesale generation tariff, comprising a capacity payment (set by the government) and an energy fee (set by market competition in the regional power markets). A formula is provided for calculating capacity payments, which include depreciation and financing costs.

Several reform proposals seek to improve the Catalogue system. However, residential pricing will remain subsidised and there is no clear principle for prices to reflect costs. The Catalogue system will be retained for end-user pricing, but the number of categories will be reduced to three: residential, agricultural and all industrial and commercial users. The first two categories will be subject to a single tariff, and the third to a two-part tariff for users with a transformer capacity of 100 kVA or more, or an overall capacity of 100 kW or more.

In addition, a range of new tariffs will be introduced including peak and off-peak, dry and wet season, high reliability and interruptible.

The average retail tariffs in a region will be built up from the average wholesale tariff, the average transmission and distribution tariff, taking into account line losses and any legitimate government levies. Tariffs for residential and agricultural users will be lower than the average, subsidised by the other users.

Retail tariffs will be reviewed and, if necessary, adjusted each year, taking account of: *(i)* the overall costs of power supply; and *(ii)* a specific mechanism that links wholesale prices (especially those relating to fuel costs) with retail prices. This linkage mechanism applies only to industrial and commercial users.

PERFORMANCE

Increasing demand puts strains on reliability

Major investments have been made in China's power sector over the last 20 years, especially in new generation capacity, reflecting China's striking development as a consumer of electric power. By 1995, China had become the world's second largest electricity consumer with a total consumption of about 1 000 TWh (Figure 6). By 2003, this had nearly doubled to 1 890 TWh. China has added, on average, more than 15 GW of new power plant capacity annually over this period.

Per capita power consumption has more than quadrupled, increasing from 340 kWh in 1983 to 1 470 kWh in 2003. Even so, per capita consumption at that time stood at only half the world average and one-eighth the level in the United States. Clearly, there is considerable room for continued rapid growth, assuming China's economy continues to expand. In 2003 and 2004, a total of nearly 75 GW new capacity was installed, and a further 66 GW was installed in 2005. More than 70 GW is expected in 2006.

Growth has continued at a striking rate. In 2004, China's power sector generated more than 2 100 TWh, 15% more than in 2003, and a record level of expansion. Installed capacity surpassed 500 GW toward the end of 2005, and the total is expected to be about 570 GW by the end of 2006. In 2005, 117 GW of new projects were approved and about 300 GW are currently under construction.

Energy savings have helped to keep growth in demand in check

More than any other major developing country, China has had notable success in saving energy. Although the power sector has grown significantly, this growth has been at a significantly lower rate than GDP growth. Against an average annual reported GDP growth of more than 9% over the past two decades, the growth rates

Figure 6 China's installed generation capacity and gross electricity generation, 1990-2005

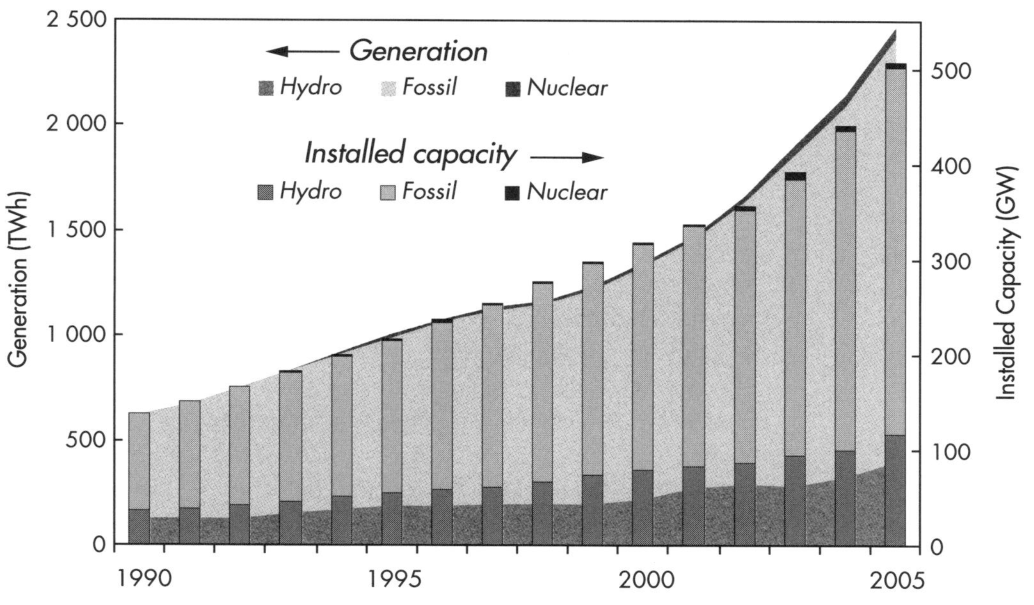

Source: National Bureau of Statistics (2001, 2004, 2005a, 2005b, 2006).

for energy and power consumption were 4.7% and 8.8%, respectively. The difference in GDP and power growth rates corresponds to an income elasticity of energy consumption[13] of roughly 0.5, and of power demand of just below 1.0. For most developing countries, energy elasticity is 1.0 or higher, and that of power, 1.5 or higher. China has, thus, offset the need to build scores of power plants and other energy infrastructure – with attendant harmful emissions. The decline in energy intensity[14] was spectacular at one stage: the absolute level of energy consumption was lower in 2001 than in 1996.

However, energy intensity remains a major concern because progress in reducing energy intensity has not been sustained. In the current phase of economic expansion, energy efficiency has declined and intensity has been rising again. China's own official statistics show a reduction in energy intensity of some 20% every 3-4 years from the early 1990s to 2001, at which point intensity flattened and began to rise. Elasticity has been well above 1.0 since then. Recent adjustments to GDP statistics and upcoming revisions to coal figures may change the details somewhat, but the overall pattern of a long-term fall in intensity followed by a more recent rise will remain the same. Despite achievements and low per capita consumption, China's energy intensity is still three times higher than that of the United States and Japan. Many industrial processes in China consume 20-40% more energy than in OECD countries. The level of final energy use efficiency is still 20% lower than that of developed countries.

The conditions that previously allowed China to reduce energy intensity included the existence of a significant proportion of production capacity that was very inefficient

13. A measure of how much energy is needed to fuel economic growth.
14. The amount of energy consumed per unit of GDP.

and was shut down. Structural reforms also reduced the role of energy-intensive industries in the economy. A different approach will be needed if energy intensity is to be brought under control again in order to meet the government's objective of reducing intensity by 20% over the next five years. Further power sector reforms need to integrate policies to contain demand and to ensure that power market players have the funds and the incentives to promote energy efficiency.

Ensuring that capacity is available, at the right time

Developments over the last decade or so suggest that China is caught in a boom-bust cycle, with recurring periods of supply shortage, as well as excess capacity. In other words, the power sector's reliability does not yet appear to be sustainable over the long term. China experienced chronic power shortages through to the late 1990s, which led to a building boom in generation. Then, for the first time since the 1980s, a temporary oversupply arose out of the general slowdown in industrial activity associated with the Asian financial crisis combined with domestic economic reforms. In 1999, the government placed a moratorium on all new power generation construction, which was only lifted in 2002. Electricity shortages again emerged in late 2002 and intensified through to 2005 (Box 5). Despite the addition of more than 50 GW of new generating capacity in 2004 alone, shortages developed in 2005, which affected 25 provinces and municipalities. The shortages led to significant economic losses from supply disruptions. They also had a major impact on the demand for oil, which is used to fuel standby electricity generators.

The cycle is now working back to an oversupply phase. A building boom in generation, sparked by the recent shortages, now threatens again to produce some overcapacity in 2007. At the same time, electricity demand growth fell to 13% in 2005 and is expected to decline further to 11% in 2006. This is the result of a gradual slowing down of economic growth, as well as a change in the nature of this growth away from energy intensive industries. In 2004 and 2005, more than 100 GW of new generating capacity was commissioned. As a consequence, the number of provinces experiencing sustained power shortages in 2006 is expected to fall considerably from the peak of 25 in 2005.

Box 5 Power supply shortages 2002-06

China has suffered periodic shortages of electrical power supply throughout its three decades of modernisation. The most significant shortages, in terms of political and economic impact, occurred between 2002-05.

Shortages started to appear in 2002 as the rate of growth of electricity demand rose to 11%, from an average of 9% in the previous two years. In that year, power shortages were reported from 12 provinces. Demand rose by 15-16% in each of the succeeding years, 2003 and 2004. As a consequence, the number of provinces suffering from sustained power shortages rose to 24 in 2004 and to 25 in 2005, despite a drop in demand growth to 13% in 2005. The shortages stemmed from the failure to construct the generating and transmission capacity needed to meet the added demand created by the energy-intensive economic boom, which was stimulated by the central government in 2002. Estimates of the capacity shortage indicate that the national deficit rose from about 10 GW in 2003, to as much as 30 GW in 2004, before declining to about 20 GW in 2005.

Location and impact of the shortages

The shortages were greatest in those regions of concentrated economic growth, namely in the East China Grid, which covers Shanghai, Jiangsu, Fujian and Anhui, and in the South China Grid, particularly in Guangdong Province. The East China Grid accounted for more than 50% of the national total deficit. The coal-rich regions of northern China, such as Shanxi and Inner Mongolia, also suffered shortages as they lacked the generating capacity to satisfy the needs of their own economies and those of demand centres to the south and east. Power shortages were exacerbated in 2003 as result of drought, particularly in provinces such as Qinghai, Gansu and Ningxia along the Yellow River. In the winter of 2003/4, an unprecedented surge in the use of electricity for household heating coincided with disruptions in the supply of coal to power stations. In most regions, the shortages translated into problems of supplying power at peak hours and were greatest in the summer months, when demand for air conditioning soared.

Measures taken

A range of measures were introduced at city and provincial levels to manage the shortages. Industrial enterprises were the usual target, and the two most common measures were planned power outages and an enforced change of operating hours. In the summer months enterprises might be restricted to operating only 3-4 days per week, or might be forced to work at night rather than during the day. In Zhejiang, power outages averaged 11 days per month in the first half of 2004. Compulsory, one-week closures were enforced in more than 6000 enterprises in Beijing each summer. Many cities saw escalators stopped, street lighting radically reduced, and the thermostats of air conditioners raised. Time-of-use pricing was introduced, or, where it was already in place, the differentials were increased.

Demand for oil

The shortage also strongly affected the demand for oil. Enterprises, and even individual householders, sought to maintain their power supplies by acquiring diesel generators. This not only led to increased business for manufacturers of generators, but also made a major contribution to the surge in China's oil import requirement in 2003 and 2004. Over these two years, demand for oil grew at an average annual rate of 13% and the level of net imports of oil more than doubled from 78 million tonnes in 2002 to 150 million tonnes in 2004. China accounted for 27% of the increase in world oil demand in 2004 (IEA estimate). Much of this rise can be attributed to the requirement for power generation, although the growing use of cars was another factor. In 2005, when oil demand for emergency power generation ebbed, net consumption of oil products rose by only about 2%, and the share of world incremental oil demand was 14%.

Reliable power as an instrument of development

China's per capita GDP has more than quadrupled relative to the 1980 level, but in purchasing power parity (PPP) terms, it remains only around one-fifth of the OECD average. At the same time, there are large income disparities across the country and the development gap is widening. There are significant inequalities between regions in terms of education, health care, economic structure, infrastructure development, FDI, and overall human development. The most developed areas of the country[15] have incomes almost double the national average. By contrast, a number of western and central area provinces still have incomes that are below those seen in low-income developing countries. The divergences between coastal provinces and the remainder of the country have been growing. Another marked feature of China's uneven development is a significant urban/rural divide in all Chinese regions.

To reduce the gap between the eastern coastal provinces and less prosperous other regions, the government has launched a vast campaign called, "Developing the West". Power sector development through grid investment is part of this strategy. The northern and western parts of the country are home to significant renewable resources, including hydro, wind and solar energy, as well as the main coal reserves. Promoting a more balanced development has implications for power sector reform, starting with the need to secure a reliable power supply that is universally accessible. Electrification is not yet complete. Over 10 million rural Chinese remain without electricity.

Pollution

Urbanisation and economic activity linked to environmental degradation

Five of the ten most polluted cities in the world are in China. Acid rain is falling on one-third of China's territory, and one-third of the urban population is breathing polluted air. Poor air quality is estimated to impose a welfare cost between 3-8% of GDP. The benefits of reducing air pollution would therefore be considerable and can be expected to exceed costs. Environmental pollution has become a growing source of social discontent, and the government recognises that the costs of neglecting the environment are increasing to unacceptable levels.

Government policies fail to contain pollution

China has partially adopted a "polluter pays" principle. The two main approaches include controls on emission quantities and fees on emissions, which have been progressively increased[16]. New policy guidelines were introduced in 2002 that require new, expanded or retrofitted coal plants to install desulphurisation equipment. New legislation, introduced in 2003, significantly increased penalties for the emission of air and water pollutants. These and earlier measures have had some effect. For example, the quantity of sulphur emissions rose only 5% between 1993 and 2003, despite GDP more than doubling in the same period. But these achievements are fragile and have been hard to sustain. Sulphur emissions have been rising again with the economic upswing. As yet, few old power plants are fitted with anti-pollution equipment and the level of pollution, both for air and water, remains high.

Power sector is a major polluter

The power industry is not the only economic sector placing enormous pressure on the environment. However, the dominance of coal in power generation and the construction of large dams mean that the environmental damage caused by the power

15. These are generally held to include the Beijing-Tianjin corridor, the Yangtze River delta and the Pearl River delta.
16. For example fees on SO_2 emissions reached USD 76 per ton in 2005.

sector in China is greater than any other industry (Box 6). Environmental pollution is, therefore, a major issue that must be factored into power sector reform plans. To date, the government has failed to effectively co-ordinate environmental and energy policies in order to tackle the issues.

Box 6

Environmental pollution from China's thermal plants

The power sector in China burns coal to generate more than three-quarters of total power supply. Low-efficiency plants, many of which have minimal or no pollution control technology, persist in this sector, although new plants are much better in terms of efficiency and particulate emissions control. The power sector used nearly 1 000 million tons of coal in 2004, and was responsible for an estimated 50% of the nation's SO_2 emissions, 80% of NO_X emissions, and 26% of CO_2 emissions. Thermal power generation, which requires large amounts of process water, is also contributing to severe water shortages and deterioration in many parts of the country. China also discharges more than 70 million tons of solid waste each year from thermal plants, which is expected to grow to about 160 million tons by 2010. Coal plants also discharge significant amounts of mercury, exposure to which can cause neurological and developmental effects in children. The health and environmental consequences include millions of cases of premature death and chronic illness each year, lowered agricultural output, and accelerated aging of infrastructure. While per capita greenhouse gas emissions are still low, the power sector is now China's largest single source of these emissions.

Source: Berrah, Lamech and Zhao (2001); The Regulatory Assistance Project (2002).

Coal plant efficiencies still well below the OECD average

The efficiency of Chinese coal plants has improved steadily, but remains low at 33% compared with a world average of 37%. Much of the improvement is the result of new, larger plants with a capacity of 300 MW or higher. If coal plant efficiencies could be further improved – and thereby offset the need for new capacity – the payoffs would be huge, both for the environment and for reliability. Every 1% improvement in overall efficiency results in coal savings of 8 million tons per year, offsetting the need for 3 GW of new capacity, and abating roughly 4 million tons of carbon emissions. If existing Chinese power plant efficiency were improved from the current 30% average to the globally recognised standard of 37%, the annual savings in CO_2 would equal approximately 80 million tons. Opportunities for improvement continue to be missed. Most of the new coal fired plants are domestically manufactured, conventional, sub-critical units. On average, these are significantly better than current units, but could have been better still. Commercially available super-critical units, which are just now being introduced in China, are about 5% more efficient than the conventional units.

Investment must increase significantly

China has been broadly successful, partly due to its reforms, in generating the investment needed to power a fast-growing economy. But this can lead to the misleading conclusion that all is well. In addition to the problem of ensuring that capacity is available at the right time, significantly more investment is needed, and critically, at a much higher annual rate than in the past (Box 7).

Box 7 China's power sector investment needs

China's provisional plans for expanding its generating capacity by 350 GW by 2010 require an additional 50-60 GW of construction each year. The IEA (2003a) envisaged a more modest rate of construction to develop an additional 860 GW from 2003 to 2030, at an average of 30 GW per year. The IEA's costing implies a total investment in generation, transmission and distribution of some USD 2000 billion over this period, or about USD 70 billion per year. Chinese estimates of unit costs lie some 20-40% lower than those used by the IEA. But given that their provisional plan for investment involves a more rapid rate of construction, it would be reasonable to infer that China will be hoping to invest some USD 50-70 billion or more per year for the foreseeable future.

The highest level of annual investment recorded in the past was about USD 34 billion (280 billion yuan) in 2003. The implied rate of investment for the coming years is double this level.

Type of investment and funding sources remain a challenge

Finding ways to finance sustained expansion will be a key challenge for China. In principle, this should not be an issue of capital availability, as it is in most other developing countries. The issue is, rather, where the investment will be sourced. China's power sector is still largely state funded, with limited private and foreign investment. China's public finances are relatively sound overall, and the state could continue to play a major role. But this is inefficient – and ultimately unsustainable.

The relative lack of foreign investment is a major issue. If foreign investors stay away, this will slow the rate at which modern management practices and technology are adopted, as well as reducing the level of competition in emerging power markets. Not least, it means that foreign investors will not fill any investment gap left by Chinese investors (state or private). As long as the future shape of power market reforms is unclear and the political need persists to keep prices low, the outlook for renewed interest from foreign investors is likely to remain poor.

There is worldwide competition for limited investor capital. At the same time, pressures are increasing for Chinese investors to seek adequate financial returns. The role of the commercial banks has increased and more power companies are seeking listings on Chinese and overseas stock markets. Companies, banks and stock markets will be seeking financial returns. The ease with which the power sector can raise funds will, to a great extent, depend on how and when the government implements pricing reforms. If the industry can make profits, the funds should flow. If uncertainty persists – or if reforms are unsuccessful – then investment may slow down.

II. NEXT STEPS

SETTING, COMMUNICATING AND IMPLEMENTING A CLEAR STRATEGY[17]

China needs to reaffirm a clear strategy for its power sector

The objectives and tasks defined in the State Council's 2002 policy document remain the government's formal baseline and have been supplemented by more detailed documents (on pricing, for example) aimed at taking the main reform goals forward. The policy document makes it clear that the strategic goal is to develop a competitive, market-based, power sector, as a means of ensuring an efficient and reliable power supply and of protecting the environment. China has implemented, decisively and effectively, important elements of its strategy. This is the case, notably, of the structural disaggregation of the industry with the emergence of a large number of generators and the judicious split of the grid into two main companies. In fact, China's efforts contrast with those of many OECD countries that have implemented a much weaker form of disaggregation. The establishment of the State Electricity Regulatory Commission underlines a commitment, at least in principle, to the importance of independent regulation. The considerable efforts directed toward establishing pilot regional power markets, with a view to gradual build up of competition in these markets, is another very positive indication of steady commitment to developing a long-term competitive market and reaping the benefits of improved efficiency and reliability.

However, some aspects of the original plan have not yet been implemented, and those that have been taken forward need more work. In addition, some issues that need attention, such as how to encourage energy efficiency alongside supply-side actions, are not covered in the existing strategy. At present, pricing reforms aimed at supporting more competitive markets and introducing greater efficiency appear to have stalled. A fragmented approach has been taken in addressing environmental issues with little, if any, attention paid to the need to integrate environmental considerations into the emerging framework for economic regulation. Structural unbundling needs to be completed, with a clean separation of generators from the grid. SERC is not yet the fully independent and empowered regulator required to oversee the new market structures. The development of competitive power markets is progressing slowly. Finally, significant delays in implementation of key reforms such as pricing suggest uncertainty as to next steps – and possibly raise questions regarding the current strategic thrust of the reform process.

The combined effect of these delays and incomplete developments is quite serious. It means that China is caught between the old planning mechanisms and a new approach. The difficulties in addressing recurrent supply shortages are symptomatic

17. Specific recommendations related to this topic are given at the end of this section.

of a policy that has, at least temporarily, lost its way. Much of the power sector remains caught in an increasingly malfunctioning planning system. These factors put China on an uncomfortable high risk/low return pathway. The risks of not having a clear governance and institutional system that works effectively, and the failure to develop more efficient pricing, are that the big problems afflicting the sector – reliability, growing energy use, pollution, and inefficient investment – will not be tackled and could even get worse.

It is important at this stage, therefore, that China reaffirms its commitment to a clearly articulated strategy. This would boost confidence among stakeholders, including investors, that there is a plan for continued reforms designed to bring greater clarity to the framework under which market players can expect to operate and to the expected changes to current arrangements, as well as commitment to make it happen. One of the most striking lessons from reform processes around the world (not just in the power sector) is the frequent failure to be clear about the strategy and to define the end point towards which the reform process is aimed. Box 8 sets out a general list of issues aimed at mitigating reform risk, which might serve as a checklist for China's policy makers. They would, of course, need to be adapted to China's specific circumstances. Some of the issues are explored in more detail later in this report.

Box 8 Lessons from other countries: managing risk in the transition

Reform means change, which entails risk even as it promises rewards. How can this risk be minimised? In the broadest terms, reform risk is about the disruption in moving from one regulatory/governance framework (A) to another (B). Many reform programmes falter because of inadequate understanding of the starting point A (the "what" of reform) and a poor definition of the end point B. They also pay insufficient attention to the process of getting from A to B (the "how"). Awareness of the many risk factors is important to eventual reform success.

Risk factors in reform

- **Incomplete definition of the starting point A.** Are all the relevant factors of the current regulatory/governance framework well understood? These can span a broad range, from fiscal systems and state ownership to relationships between different parts of government.

- **Incomplete definition of the end point B.** Is there a clear vision of the strategic objective? There may be too many objectives, or they may not be clearly articulated.

- **Different types of risk.** Reform often starts with a simple economic objective at its core: how to make a sector or an economy more efficient. Economic change may, however, carry risks to health and safety, security of energy supply, social welfare, equity, or environmental objectives.

- **Not all eventualities can be predicted.** Reform is often "learning by doing". The broader consequences of a specific reform cannot always be predicted in advance. For example, contracting out public services may ease fiscal constraints but subsequent poor performance may cause other problems.

- **The long haul of reform.** Effective and durable reform is a long-term and dynamic process, not a one-off event that can be quickly brought to a close.

- **Opposition to reform.** Reform should improve resource allocation, which requires reducing or taking away rents and/or acquired rights. In the long run, economic performance improves and everyone benefits, but the losers are usually more sharply aware of their immediate losses than beneficiaries are of diffuse, long-term benefits.

- **Changes in responsibilities.** Pre-reform responsibilities are often centralised and reasonably clear. The change to more dynamic and market-oriented frameworks tends to disperse responsibilities in some sectors, such as the infrastructure sectors. There is also a need to maintain the right balance between central control and delegated responsibility.

- **Inappropriate rules for a new market/governance framework.** Rules that work under one framework are unlikely to be as effective under another.

- **Specific risks tied to particular sectors.** Systemic risk, *i.e.* a failure in one part of the value chain that affects the performance of other parts of the value chain, is normally associated with banking, but it poses a danger in the electricity and other network sectors.

Managing reform risk

Policies and strategies

- **Communication.** Effective communication of benefits, risks and schedules during reform helps to build enthusiasm, provides reassurance, and promotes frankness. A communication plan, with explanations suited to different audiences (technical stakeholders, the general public, etc.) is helpful.

- **Reform strategy.** A clear and comprehensive strategy sets a vision of the end point B and identifies the essential steps and activities for getting there.

- **Reform sequencing.** There are limits to what a country or society can undertake simultaneously. This includes social aspects such as subsidised pricing of services: perhaps subsidies need to be unwound slowly (but surely). There are also practical considerations of what comes first. Structural separation should probably come before competition in infrastructure sectors with a natural monopoly core.

- **Building advocates for reform.** Fostering a network of advocates, for example *via ad hoc* task forces, helps to spread ownership and enhance communication. Business is often a natural ally of reform (though beware of vested interests). Consumers can also lend their voice.

- **Defusing opponents of reform.** It is important to engage the vested interests that stand to lose in reform. They can be sectoral (*e.g.* infrastructure sectors) or regional (subsidised for public services in the past). Failure to undertake substantial action to compensate losers may undermine reform. An active and honest approach is better than pretending the problem does not exist.

- **Using external levers for reform.** Market openness and membership in a regional grouping are tested factors of success in economic reform.

- **Ensuring (if at all possible) early results in visible areas.** This helps to sustain enthusiasm for ongoing reform and helps reform to acquire its own positive momentum.

Capacities and institutions

- **Identifying, using, and possibly setting up well-functioning institutions that can help move the reform process forward.** A national audit office, competition authority or other existing or new part of the government can help to co-ordinate, monitor, and encourage reform. Networks of new and existing pro-reform institutions can help produce "champions" of reform. Spreading ownership of reform across many stakeholders ensures that reform champions emerge who will outlast the demise of any particular supporter. Business and consumer groups can also serve as champions.

- **Supportive finance ministries.** Many, if not most, reforms have a financial dimension, and they may be driven by a budget crisis or fiscal constraints. Finance ministries can be very important in lending weight to reform, and can play a valuable advocacy role within government.

- **Managing the different levels of government.** Reform needs to percolate through all levels of government. Reform at the centre of government will be undermined if it is not picked up by other levels.

- **An adaptable bureaucracy.** Civil servants play a pivotal role in reform. They may feel threatened by it, and can slow or block it. The process of culture change takes time but can be reinforced in many different ways, *e.g.* incentives, performance pay, new contracts, training, and opening some posts to outsiders.

- **Consultation and feedback.** It is important to ensure that those responsible for implementing specific changes are directly involved, so that they can comment on feasibility and risks.

- **International regulatory clubs and exchanges.** This includes sharing information on regulatory developments and experiments, keeping up with best practice and joining regulatory clubs.

Tools and rules

- **Targets, controls and evaluation.** Measurable targets should be set, strong monitoring and evaluation processes should be established, and incentives to meet the targets should be put in place. Publicising the results can help to sustain reform momentum.

- **Compliance and enforcement.** This starts with the design of rules that are likely to generate a high degree of compliance (stakeholder consultation can help). It also implies ensuring that enforcement powers are adequate.

- **Defining new rules for new market and governance frameworks.** Beware of deregulation. In many (not all) areas, the objective is to change the rules, not remove them entirely. This applies especially to the infrastructure sectors.

- **Clearing out obstructive rules.** This applies to rules that are inappropriate or even positively unhelpful.

- **Allocating responsibilities under new market and governance frameworks.** Responsibilities for control, monitoring, security, etc. need to be clearly allocated between government and market players, or between different parts of government.

- **Tools for specific purposes.** Tools have been developed to manage markets that include a mix of public and private enterprises: competitive neutrality frameworks, corporate governance frameworks, fiscal unbundling, etc.

- **Carrots and sticks (rewards and punishments).** These may take the shape of fiscal incentives, for example, between levels of government.

- **Devices to lock in the reform process.** This may be a public (and well-publicised) agreement that includes clear benchmarks for progress.

Reaffirming a clear strategy is, at the same time, an opportunity for China to review – and improve upon – the 2002 strategy. The most important objective of the review would include the following:

- *Integrate energy-efficiency goals and measures to achieve these alongside existing supply-side goals.* In order to align China's power sector reforms (both now and for the longer term) with the country's overall development strategy, policies to reduce demand should be given as much consideration as those aimed at strengthening investment in generation and the grid. A two-pronged approach has the best prospects for mitigating (if not eliminating) China's uncomfortable boom/bust cycle of supply/demand imbalances.

- *Ensure that the emerging regulatory framework includes incentives to improve environmental performance.* Actions taken so far appear to segregate environmental policies from the economic regulation of the power sector, limiting the overall effectiveness of regulation. The new regulatory framework must give all participants a clear view of costs, including environmental costs. Incentives to produce cleaner power based on the "polluter pays" principle (taxes, subsidies, obligations or quotas, etc.) must be underpinned by better collection and communication of emissions data alongside other information.

- *Distinguish between near-term and longer-term actions to improve performance.* China needs to devote effort now to reform activities that can yield positive near-term benefits while also helping to lay the groundwork for fully competitive markets. These include: strengthening the institutional framework; integrating energy efficiency and environmental objectives more firmly into current regulation and future reform plans; and implementing pricing reforms to support improved economic and energy efficiency. Taking actions, including modest steps towards competitive trading, to establish a sounder groundwork for the efficient development of fully competitive power markets in due course is also important at this stage.

Establishing a strong legal foundation

A revised framework for electricity law is an essential starting point for capturing and laying out the reform strategy, and for providing the legal foundations for both near- and longer-term reforms (Box 9).

The current framework for electricity law was passed in 1995, during a period of severe power shortages, to encourage the construction of new plants. It has since been overtaken by reform, and does not provide an adequate legal basis for reform initiatives such as the introduction of direct transactions between generators and customers, the development of consumer choice of supplier, and a new approach to grid tarification. Despite this disconnect between old legislation and a new situation, the law remains the main instrument governing the operation of the power system. SERC is required, by the State Council executive order under which it was set up, to provide amendments to the 1995 law and a revised law is currently under discussion. This is an important opportunity to set out a clear framework for further reforms and to encourage policy and regulatory stability. It would be especially helpful in supporting the development of SERC as the market regulator[18]. Establishing a strong central law may also encourage local level agencies to more clearly reflect its main instruments for effective regulation within their own laws and implementation rules[19].

18. If it takes too much time to agree to a revised law, or if the law looks inadequate, then the advice of an Energy Foundation report (2002) remains appropriate. "We suggest that administrative regulations be authorised as legal by the government ahead of formally legislated laws, because only when the regulatory agency and its functions are authorised by law, can a new transparent, predictable and balanced system be developed". See also Berrah and Wright (2002).

19. Australia's National Electricity Law is an interesting example of a "cascade" effect at work. It has been passed in practice by state parliaments agreeing on one state taking the lead, after which mirror legislation has been quickly passed in the other states.

Box 9 Lessons from other countries: establishing a strong electricity law

A clear legislative framework underpins power market reform in nearly all OECD countries that have taken on this challenge (Green, 1997; The Regulatory Assistance Project, 2000; IEA, 2001). In contrast with legislation initially established for a state-owned power system – when the focus was typically on technical and safety issues – many vastly different factors must be considered prior to market opening. Key issues for developing new law that effectively supports a competitive market are:

- **Coverage.** The law should set out clearly the strategic objectives for the power sector. This is an opportunity to emphasise a new, holistic approach that incorporates energy efficiency and environmental considerations which need to be woven into the new regulatory framework. This is, of course, in addition to the more "classic" issues related to economic regulation of the sector, which include: the regulator's mandate and powers; the establishment and governance of an independent system operator; the methodology for transmission pricing; and provision, in due course, for the extension of consumer choice of suppliers.

- **Secondary rules.** Some issues are best developed outside the main law – as long as the main law is clear on the strategic fundamentals and the process for developing secondary rules is robust.

- **Flexibility for change and review.** Power markets evolve and so do regulatory needs. It is important to avoid a situation where it is impossible to adapt legislation. Mechanisms to support flexibility should be built into legislation with the anticipation that new laws may need to be developed in a few years' time.

A reform champion and broader leadership are needed to sustain momentum

Reform processes in other parts of the world point clearly to the importance of having a strong and stable policy advocate (or advocates) for reform in the centre of government. This role is different from, though complementary to, the role of regulator. Although integration with broader energy policy is important, the power sector raises too many complex issues for the reform process to be handled exclusively through a broad energy agency approach. A champion can help to minimise uncertainty, maintain coherence and ensure stability, thereby building confidence among market players. In the absence of an energy ministry, the current focal point is the NDRC, which co-operates to some extent with SERC. The development of a new energy ministry or agency is clearly relevant, but the important issue is to ensure a clear central point for the power sector within the institutional framework for the entire energy sector. Authority and seniority, together with an adequate staff and budget[20] are prerequisites of any agency that fills this role.

If the NDRC is to remain the focal point, it will be necessary to distinguish more clearly between policy and regulatory responsibilities, which are currently shared between NDRC and SERC, as these two agencies currently overlap in key areas of

20. NRDC's energy bureau is very understaffed, even with the recent increase of staff from about 30 to 50, and even accounting for the energy administrators in corresponding positions in provincial and municipal governments.

policy development. For example, SERC is tasked to organise the implementation of the power sector reform plan and to provide recommendations on further reform: these are policy rather than regulatory functions. Distinguishing between policy and regulatory functions is not an exact science, and it may be that at this early stage of power market reform – and in the absence of a revised electricity law that sets out a clear framework – both SERC and NDRC need to be involved. Policy development is a grey zone, to which the regulator needs to contribute. Day-to-day regulation highlights issues within the regulatory framework and helps to identify policy directions that need corrective action; thus, the regulator's input should be taken seriously. But it is best that policy comes to rest firmly with a separate entity to allow the regulator to focus on the task of implementing the established framework. This division of labour also leaves the policy maker free to manage broader responsibilities for issues such as security of energy supply, which involves co-ordination with other energy colleagues.

A related priority is to tighten co-ordination among the various sections of government that play a role in power sector reform. The Chinese approach of setting up a "Leading Group" may be effective. These groups create bridges between leaders at the apex of the political system, thereby contributing to the coherence and co-ordination of policy decisions. A number of other government agencies are potentially relevant to electricity reform, including the Ministry of Finance, the Ministry of Land and Resources, the Ministry of Science and Technology, the Ministry of Commerce, and the State Environmental Protection Administration (SEPA). Involving environmental and health officials (including individuals from the NDRC's own environment bureau) in policy discussions on power reform is especially important at this stage. It is critically important to give early consideration to the environmentally damaging effects of current power production. A widely based group of stakeholders would help sustain coherence and understanding of often complex issues. It may also help to prevent the "hijack" of the reform debate by minority vested interests.

Responsibilities of the reform champion

The reform champion has two main roles: dealing with vested interests and promoting reform "ownership". The former is probably the biggest challenge. Reform can be expected to create winners and losers. Stakeholders that perceive themselves as losers may try to influence implementation in a way that seriously distorts the substance or timing of original plans. The aim for the champion is to manage transition in a way that secures support from key stakeholders, balances competing interests and maintains the essential integrity of reform. It is important for the champion to distinguish between narrow vested interests (those who stand to lose their profits) and broader losers (poor people, unless steps are taken to establish lifeline support, which is discussed later).

As the reform process is rolled out, the champion should also hold responsibility for adjusting the policy programme to fit developing circumstances, without losing sight of the main strategy. Power sector reform is not a static or "one-off" process. It is hard to predict how details will roll out. A constant evolution must be expected, which calls for vigilant monitoring and evaluation. Unanticipated technical issues will arise in relation to market design, market structure and regulation. For example, the reorganisation of Nordpool in 2002, and the Australian review and subsequent changes to strengthen NEM in the same year, were carried out to update the respective frameworks in the light of market developments.

One of the most effective means of creating a sense of ownership is to encourage consumer participation in the reform process. Full consumer choice is not a realistic or helpful option at this stage of China's power sector reforms. But enhancing consumer participation in the regulatory process to ensure that their needs are taken into account is both feasible and helpful. Lack of customer orientation and poor consideration of consumer views have been a weakness of most pre-reform supply-biased frameworks and state-owned monopolies. This mindset continues today, to some extent. Regulators tend to focus on supply-side generation, system operation, and grid tarification while neglecting consumer aspects, although some countries' reforms have included consumer protection measures such as complaint-resolution mechanisms.

Consumers in some developing countries, such as Latin America, have recently started to complain about their lack of representation in regulatory processes, tariff decisions and dispute resolution[21]. A strong consumer voice is desirable for a number of reasons. For instance, it counterbalances vested interests, such as the strong influence usually exerted by the industry. It also provides regulators with political support and protects them from undue political interference in their rule making. Consumer participation can be especially valuable in the process of tariff rebalancing as it can enhance credibility and may make the outcome more acceptable.

A strategic focus is absolutely essential to reform. But it must also be backed up by the translation of strategic objectives into detailed, practical implementation plans and actions. The champion can further build ownership and stakeholder confidence by ensuring timely and effective implementation of reform activities. China seems to experience consistent difficulties in this area. For example, the first pricing reforms were announced two years ago, but have proceeded very slowly. Uncertainty and delays for key reform elements, such as pricing, weaken the intent and hence undermine confidence that reforms will be implemented as planned. China needs to ensure that implementation details are well prepared in advance and avoid delays once a specific reform is announced. When preparing a reform, care must be taken either to include implementation arrangements or else to specify a transparent mechanism by which implementation rules will be designed. Timelines should be fixed, communicated, and adhered to.

Ultimately, reform requires new approaches to compliance and enforcement. This is currently a weak aspect of China's power sector governance, which is linked to the power and control of local governments, as well as the lack of a clear and strong legal framework to underpin power sector regulation. The aim should be to create a situation in which compliance makes good business sense and becomes voluntary. The use of sanctions as a deterrent is only one mechanism that contributes to creating compliance. Others are the quality of regulations of the rule-making process, the

21. Apoyo Consultoria (2002); World Bank (2004).

opportunity for public participation in the process and in the application and enforcement of rules, and overall access to information. Voluntary compliance is linked to sound regulatory design. Compliance problems usually reflect two things: limited acceptance of the new rules of the game and consequent tensions between interests. In fact, a regulation should not be adopted if compliance prospects are poor. The OECD (2005a) points out that solutions do not lie in occasional enforcement campaigns; these have little effect on long-term behaviour.

Finally, the reform champion should be given the task of establishing a mechanism for monitoring and evaluation. The goal is to assess the measures of success and the symptoms of failure as each reform is implemented. Given the complexity of China's reform situation – and the delicate process of moving forward in stages, according to a clear overall strategy – it is essential that the government be able to recognise and address these signals as further reforms roll out. This can be done through careful monitoring of key indicators such as:

- **Effects on investment.** Successful reform is typically marked by continuing investment in all parts of the industry, which ideally strikes a balance between supply- and demand-side energy-efficiency investments, and promotes cleaner energy.
- **Emerging competition in wholesale markets.** For example, within the regional power markets, success will be marked by ensuring that no single generating company gains excessive market power and that players cannot collude in price setting.
- **Effects of reform on the end user.** This includes quality of service and continuity of supply (reliability).
- **The effectiveness of incentives** for clean energy production and renewable energy.

The need for transparency

China has some way to go in achieving the transparency levels required for an effective reform process and for the subsequent efficient operation of its power markets. At present, relevant information is not always made public. Thus, few stakeholders understand current pricing methods and the complex fee structures. It is also hard to judge current levels of profit, subsidies and cross subsidisation. Opportunities to participate in the regulatory process are unclear, as are right to appeal against regulatory decisions. Indeed, it is possible that the reforms to date have actually complicated matters, at least as regards generators. Their numbers have multiplied, but many retain strong and often opaque links with state funds. At this stage, greater transparency is especially important in relation to public funding, costs and prices, as well as clarifying institutional powers and responsibilities. It could be expected to improve both the quality of the regulatory process and the efficiency of new power markets. It may also help to more effectively manage the relationship between the centre and local levels of government. Transparency has already proven its value in one area: pressures exerted by a better-informed public have forced the government to take action to mitigate pollution and increase the profile of environmental objectives.

Effective communication of electricity reform is complex, sensitive, takes time, and requires forward thinking and – indeed – an effective communication strategy. China

needs to make its intentions clear to stakeholders, in ways that address the needs of specific target groups. The details, which are often technical and difficult to convey, are important for certain stakeholders such as market players. It is equally important to provide users and the wider public with the broad picture, including the overall objectives and the expected results. Box 10 below sets out some general considerations related to transparency, which may help China's efforts to identify the most effective approach within its own governance context, as well as what can be realistically implemented.

One essential element of the framework for transparency is more open exchange of information, particularly reliable data on the power sector and supply/demand developments. China already collects and publishes a great deal of information, but more, and more timely information will help lay the foundations for better performance across all classes of generators and consumers. The data needed for effective regulation of the power sector often differs from information gathered for other statutory purposes (Box 10).

Box 10 Lesson from other countries: transparency is a core feature of successful reform

What is transparency?

In the context of power market reform, transparency is – first and foremost – an issue of being able to access information. One broad definition is the ability of businesses to understand fully the regulatory environment in which they operate. Clear understanding improves the quality of market access and competition. A core transparency requirement is that all stakeholders should know what the rules are, where to find them, how they are applied and the penalties for non-compliance. This applies to existing rules, as well as to information on prospective rules.

Transparency also implies opportunity for consultation. Giving stakeholders a voice in regulatory decision-making helps to strengthen the quality of the rules that emerge and enhances the likelihood of compliance. More open and accessible procedures are more legitimate, less vulnerable to capture, and more likely to bring high quality information that improves analysis of regulatory policy options. The experience of OECD member countries shows that an open and transparent consultation process for developing new rules helps to ensure that these are robust and "fit for purpose" (IEA, 2005a).

Why is it so important?

Transparency is important for a number of reasons:

- It helps to avoid regulatory capture and prevent corruption in the regulatory process. Market players (consumers, investors, companies, etc.) can apply pressure for rules to be applied as intended.

- It supports the provision of essential information – costs, prices, etc. – without which competitive power markets cannot function, or function inefficiently.
- It ensures that rules well designed, *i.e.* neither more nor less than is needed, which promotes compliance.
- It identifies who holds responsibility for each aspect of the new market framework.
- It shows where a regulator's powers are relatively weak, making its views – and the responses of regulated firms and other key players – a matter of public record to help strengthen its impact.
- It reveals information about market rules supporting a range of other specific purposes such as the prevention of cross-subsidisation between the competitive and monopoly elements of the value chain.
- It provides regular communication about reform developments, which can help to sustain reform enthusiasm.

How can it be secured?

A number of mechanisms to support transparency exist. A clear legal and regulatory framework is the essential starting point, including clearly written and unambiguous rules. Other mechanisms include:

- Posting information in the public domain (*e.g.* on Web sites), providing access to consolidated databases, establishing enquiry points.
- Benchmarking, both in relation to regulatory approaches and market developments. Careful observation of how the regional markets evolve. Systematic monitoring and data collection. A good example is the annual EU benchmarking report on progress with the development of the Single Market in gas and electricity. International benchmarking can also be a useful way of checking progress against international best practice, for example power sector performance standards (outages, supply disruptions, etc.).
- Consultative regulatory decision-making processes, such as calls for comments on proposed new rules.

Recommendations for reaffirming a clear strategy

Reaffirm long-term strategy; undertake near-term actions. China should reaffirm and update its strategy for further power sector reforms, thereby reiterating its commitment to the long-term development of competitive power markets. The strategy should explicitly integrate energy-efficiency goals and environmental objectives and should distinguish between near-term and longer-term issues, and specify related actions. Near-term priorities should include efforts to:

- Strengthen the institutional framework.
- Integrate energy efficiency and environmental objectives into current regulation and future reform plans.
- Implement pricing reforms to support improved economic and energy efficiency.
- Establish a solid foundation for developing competitive power markets, including a first set of measures to stimulate basic competitive trading across China's regions.

Update electricity law. China should move swiftly to update its electricity law to provide a firm foundation for further reforms and to clearly establish the strategic objectives for the power sector. This is an opportunity to emphasise a new holistic approach for the regulatory framework that:

- Incorporates energy efficiency and environmental considerations, as well as economic regulation of the sector.
- Confirms SERC as an independent regulator.
- Allows for adaptation as the power sector evolves, to avoid the need for further revisions in the near future.

Appoint a reform champion. China should identify a focal point within the institutional structure that will act as the "champion" for taking policy reforms forward. The champion can play a key role in clarifying the separation between policy and regulatory functions; it needs to be adequately resourced to:

- Provide further policy support for reform.
- Establish a mechanism to reactivate momentum for the reform process within government.
- Create a "Leading Group" to tighten co-ordination on electricity issues between different parts of government.

Without losing sight of the main overall strategy, the champion should be prepared to carry out many specific tasks, such as:

- Dealing with vested interests.
- Promoting reform "ownership".
- Adjusting the policy programme to fit developing circumstances.
- Encouraging consumer participation in the reform process.
- Establishing an implementation plan, and ensuring that it is met.

- Encouraging new approaches to compliance and enforcement.

- Establishing a mechanism for monitoring and evaluation.

Increase transparency, improve communication. China should take steps to increase transparency of the reform process, and should improve communication of its strategy and specific reform plans. Relevant information should be publicised and disseminated to appropriate target audiences:

- Opportunities should be created for stakeholders to participate in the reform process, for example by commenting on specific proposals.

- The broad picture, objectives and expected results should be explained to users and the wider public.

- More technical information should be made available to market players.

Improve data collection, analysis and access. China should find ways to improve its systems for data collection and analysis on the power sector, and provide stakeholders with better access to information. This is an essential element of transparency and will increase stakeholder understanding of power sector developments. SERC should play a leading role in this process, in order to avoid potential distortions by parties with a vested interest. The institutional design should ensure long-term support for collection of statistics that cover all relevant aspects of the electricity system. Key information that needs to be available includes:

- Current demand and demand peaks.

- Projections of future supply and demand by region.

- Clear and detailed knowledge of the grid and generation infrastructure capacities.

STRENGTHENING INSTITUTIONAL CAPACITIES[22]

Broad governance challenges on the path to further reform need to be taken into account

China faces some important challenges on the path to further power sector reform that cannot all be resolved by policies aimed at the power sector alone. These underline the need for caution in moving forward too quickly with competitive power markets. China's accession to the WTO in 2001 has accelerated change towards a socialist market economy, but it is a slow process. Meanwhile, awareness of the weak spots in the broader framework may help the development of strategies to manage the weaknesses. It may be possible to strengthen capacities in some areas, such as monitoring anti-competitive behaviour in power markets, ahead of broader governance progress, such as the development of a competition authority.

22. Specific recommendations related to this topic are given at the end of this section.

China's governance traditionally does not make a clean separation between political and economic decisions, *i.e.* between the government and the market. Linked to this, there is no tradition of true independence between the various organs of government. The executive (headed by the State Council) is largely dependent on the Communist Party. The legislature (the National People's Congress) and the judiciary have little real independence. To take account of China's unique features, it will be necessary to adapt the regulatory models that work in other countries, at least for a transitional phase. It may not be possible, for example, to move directly to a fully independent regulator, but core functions and features of effective regulation can – and should – still be developed.

To understand the change that is taking place in China's power market, there is a need to appreciate how it was structured before. It has been said that "(China's) governance system was quite simple". The planning system tended to limit "the world of the possible", and thus the scope for officials to exert power. The power structure was highly centralised; local governments, enterprises, institutions and social organisations had no autonomy and were *de facto* branches of the government in Beijing. They executed the Plan under the guidance of the State Planning Commission (now NDRC). All economic activities were tightly controlled. The role of China's government in the past was to plan and direct investment, production and consumption, primarily through the use of quotas and fixed prices. Central government also controlled the distribution of human and financial resources. Relationships between the different levels of government were based on the need to formulate and implement the plans.

Both planning and regulation for a competitive market require active engagement by government, but there is a sharp difference between the two. The planning tradition, based on top-down command and control, does not take costs as its starting point. Thus, it bears little relation to regulation, which requires equally active but more arm's length and incentive-based management of horizontal market relationships, as well as methodologies that emphasise the need to reflect *all* costs in pricing and investments across the power sector value chain. In the absence of a restraining regulator, the relaxing of the old planning system alongside with monetisation, has increased the scope for corruption. To use a Chinese idiom, the current management of the power sector can be described as "tumbling over the threshold, both feet not on the ground".

Part of China's current problem within the reform process is that companies that have been "set free" quickly learnt how to benefit from a situation in which they are no longer part of "a plan" but are not yet under the effective supervision of a regulator. In fact, China's companies are exhibiting the typical behaviour found in open markets around the globe. They are not interested in competition, but rather, in destroying their competitors (if the situation threatens the rent they can extract) or else in colluding. They find it more lucrative to grow market share and increase profits, and far less interesting to cut costs and become more efficient. This type of company behaviour is not specific to China, and can be expected. The main difference between China and those (relatively few) countries that have successfully reformed their power sector is the capacity to restrain behaviour through an effective institutional and legal framework (typically, a regulator and the competition authority, and more broadly, the judiciary).

Vis-à-vis other consumer products, electricity has unique characteristics – *i.e.* the need to match supply with demand at all times across the grid and a natural monopoly at the core of the power sector – and corresponding challenges. In order to effectively manage the interface between the central monopoly and the competitive components of the market, the power sector depends heavily on complex governance and regulatory frameworks.

It is well documented that the weakness or absence of important institutional and legal features has a negative impact on market economies. Three aspects that are especially important for power sector reform stand out:

- **The law and the law-making process.** A robust legal framework is vitally important for power sector reform. China is still in transition from governing by executive order to the rule of law. Despite improvements, the law is still broadly viewed by government as an instrument of policy and control of the economy and society. For example, administrative approvals are still used as a form of state control on economic activities and little emphasis is placed on protecting market processes or on controlling the power of the state. The body of law generated through the law-making process is inconsistent and ambiguous, reflecting the involvement of a large number of actors with powers to draft primary laws, as well as to adopt subordinate and implementing legislation once a law is adopted. To complicate matters further, these actors often have conflicting interests. The current ambiguity reflects the need to broker compromises between competing agencies. It also highlights the need to address the power that the ambiguity grants to bureaucrats, particularly at the stage of implementation.

- **Enforcement of the law.** Compliance with, and enforcement of, rules are major issues for the power sector and the effective roll out of reform. This links to the judiciary, which is currently weak, lacking independence and insufficiently transparent: characteristics that have a negative effect on enforcement. In reality, the judiciary does not yet have the capacity to provide adequate oversight of regulatory decisions or to effectively deal with appeals. These weaknesses, combined with ambiguity of current laws and regulations, undermine the predictability of the legal process. A related factor is that courts have yet to gain significant authority and enforcement powers remain largely with the executive of the Communist Party (the State Council, as well as various commissions and ministries, and a range of non-state actors at both national and local levels).

- **Competition policy and law.** Managing competition in reformed power markets is another priority. Strong anti-trust laws do not yet exist, which raises a number of challenges. There is an urgent need to create a competition authority (though this would not serve only the power sector).

The issue of local protectionism

Local protectionism is another key issue for reform in China. The need to fund social and economic development puts pressure on local governments to act in a variety of ways that work against well-functioning markets. Such local protectionism prevents the flow of power between some regions in order to preserve jobs and maintain tax revenues. China is not unique in needing to face this issue, but the challenges are even greater due to the size of the country. In such a large jurisdiction, central government control over local economies creates a number of issues, including the difficulty of reconciling long-term national goals with short-term local interests. In reality, centrally developed and adopted laws and regulations are not applied in a systematic way (or sometimes not applied at all). Corruption is a major issue, as is the proliferation of illegal taxes and fees, which are a burden on local generators, as well as some consumers. Symptoms of these problems in the power sector include the unregulated construction of new power plants and the protection of local power generators from outside competition.

China's power industry has its roots in the provinces rather than at national level. The unbundling of the sector that has taken place over the last 20 years has, unwittingly, encouraged the re-emergence of local control.

A lasting solution to local protectionism will require major reforms that are not specific to the power sector. For example, fiscal reforms will be needed to tackle the shortage of local revenues to meet decentralised responsibilities. This involves the very difficult problem of separating local governments from enterprises, including ordering officials to sell shares (a measure that was already implemented in the coal mining sector).

Reforms to date fall short of establishing independent regulation

To date, China's reforms have focused primarily on disaggregating the industry from government functions. The challenge now lies in developing a robust framework within government, at all levels, to regulate this "new" industry.

The establishment of SERC in 2002 was a clear signal of China's commitment to establish independent regulation. Despite this and other encouraging developments, the sector does not yet have a recognisable independent regulator. SERC remains weak, both in terms of powers and resources. Its funding is not clearly defined and some of the staff is reportedly "donated" by the power companies. Moreover, its independence from the NDRC is not yet evident. The fact that SERC reports directly to the State Council makes it seem independent from the NDRC, but genuine independence would require a clear separation of functions. In reality, policy and regulatory functions remain muddled between the two entities. Thus, SERC's independence *vis-à-vis* both the government and the regulated companies is questionable.

The widely accepted concept[23] of an independent regulator – *i.e.* a regulator that has the powers, status, and capacity to act independently of the rest of the government (albeit with provisions for accountability) – remains a challenge to China in its current phase of market economy development. It contradicts the tradition, under China's political system, that no part of government is independent of other parts. This has a direct, practical bearing on the prospects for setting up an effective regulator.

23. This is a common understanding, at least in principle. However, many countries do not yet have a genuinely independent regulator that fits this description.

It is not a question of abandoning the goal of independent regulation, but rather recognition of the need to take an approach that incrementally fosters a new, independent regulatory culture[24]. This should be supported by specific arrangements to promote a gradual separation between SERC and NDRC. Other countries that faced similar challenges because their institutional structures were not "right" for supporting independent regulators show that, given the right conditions, such agencies can quickly establish themselves. Many have undertaken incremental steps such as those China must now pursue for SERC including: putting in place separate and transparent recruitment and selection procedures, and conditions of office; establishing separate financing, preferably through a levy on the industry; and creating opportunities for training and regular contact with groups of overseas regulators (*e.g.* with one of the European clubs[25], or with other regulators in Asia).

It is important to emphasise that the transitional approach does not represent a weak option. Rather it reflects the need to co-ordinate these initiatives with others that seek to address two additional important areas which need further development: defining SERC's objectives, responsibilities and powers and establishing its institutional features.

Clarifying SERC's regulatory objectives, responsibilities and powers of enforcement

The need to focus on key near-term objectives and responsibilities

Establishing a completely independent regulator is, in reality, a longer-term goal for China. However, it is vitally important to pursue certain objectives and responsibilities in the near term. Others can be developed or given greater priority at a later time, as the reform process gathers pace and as the need arises. The most important aspect at this stage is the development of strategic objectives, which is also an opportunity to complete and clarify the unbundling of policy and regulatory functions *vis-à-vis* the NDRC. The objectives and responsibilities listed below are all relevant to SERC, at this point in the reform process. Some are already in its remit, but may need strengthening:

- **Pricing.** At the very least, SERC must have a strong say in the development of pricing policy and hold ultimate responsibility for enforcing pricing regulation (this issue is considered in more detail below).

- **Regulatory oversight of current market players.** SERC should be responsible for overseeing actions to strengthen the corporate governance of power market players (including generators and the grid companies) and to police separation of generation assets from other state interests. It should also manage licensing of generators.

- **Regulation of system dispatch.** An independent regulator is necessary to ensure that system dispatch is fair. SERC should also oversee the development of improved system security, and of pilot regional power markets (as now) and competitive trading.

- **Promotion and implementation of energy-efficiency policies.** SERC should take an active role in this area via adjustments to the economic regulatory framework.

24. France may provide a useful example of a regulator which has evolved by stages, and is well set up with clear responsibilities and resources that enable these responsibilities to be carried out.
25. The Council of European Energy Regulators (CEER) is a possibility. CEER was established as a forum for debate among EU gas and electricity regulators.

- **Development of policies for the environment.** SERC should take an active role in policy development and in ensuring that new policies are properly translated into the economic regulatory framework (*e.g.* as regards investment planning). This is particularly important given the anticipated heavy reliance on coal.

- **Approval of generation and grid investment plans.** SERC should gradually become more involved in applying project approval criteria that have been determined by planning authorities.

- **Data collection and analysis on the power sector.** SERC must take steps to avoid being dependent on others (such as the grid companies) for essential information that improves understanding of power sector developments (including current demand and demand peaks, projections of future supply and demand by region, and a clear and detailed knowledge of grid and generation infrastructure capacities).

These responsibilities must be supported with appropriate enforcement powers. These include powers to require regulated entities to report information in accordance with statutes, as well as powers to resolve disputes in the market by acting as referee for competition issues (discussed in more detail below).

Responsibility for pricing

Pricing is a controversial issue in the China context. It is a function that would normally be held by the regulator, but which is currently with the NRDC. Effective power price regulation may be possible only when the regulator is in charge, so that decisions are not influenced by political considerations. At the same time, the regulator needs to make its decisions within a framework that is approved through political and law-making processes. One approach might be:

- The primary law, passed by NPC, establishes the methodology for law-making, building in flexibility for adjustments. This would ensure that primary law does not need to be revised in the light of experience as the reform process rolls out.

- If necessary, NDRC would oversee changes to the methodology, with advice from SERC. All such changes would be made public.

- SERC implements the legally mandated methodology. It monitors costs, has authority to take action against abuses, and enforces the rules. If there are any disagreements, SERC's advice is made public.

Monitoring and managing competition issues

In the absence of a general competition authority[26], and pending the broader development of competition law and institutions in China, SERC will need to take on the additional responsibility of monitoring and managing competition issues. In other countries, regulators and competition authorities play a very active role in the promotion of competition and the identification and prevention of anti-competitive behaviour (Box 11). This is one way of avoiding situations in which existing

26. Further down the road for China, there will be the need to define respective responsibilities with the competition authority, and set up co-ordination mechanisms. A wide range of mechanisms has been put in place around the OECD. The three main approaches are for the regulator *(i)* to share competition powers with the competition authority, as in the United Kingdom, *(ii)* to have its own powers which are separate from the competition law, or *(iii)* to form part of sector specific arrangements within the competition institutional framework, as in Australia.

incumbents retain or further develop market power, or a new monopolist or an oligopoly emerges, which might undermine the intended gains of lower costs, higher productivity and lower prices. China is, in fact, already caught in this situation: development of new entities is largely unregulated and often under the "protection" of local governments. In addition, the five generators that were unbundled from the grid have already expanded their market shares.

Power markets raise special challenges and competition agencies tend to develop considerable expertise in key areas such as: appraising the structure, behaviour and performance of markets; identifying abuses of market power; and developing remedies. These remedies might include divestment of assets, prohibition tactics such as predatory pricing and exclusive dealing, and blocking mergers and acquisitions that may damage competition. In most cases, the competition authority and the regulator would collaborate to deal with such issues. Absent a competition authority, SERC will need to take on a stronger role than is usual for the regulator, and rapidly develop its capacities to manage competition issues in the new power markets. China has recently passed an anti-monopoly law, which provides an opportunity to tackle some of these issues, at least in a general way.

Box 11 Lessons from other countries: rapid development of independent regulators key to market development

Electricity regulators have undergone a rapid development over the past few years, and their responsibilities appear to be converging. This is illustrated in a 2004 report for the Council of European Energy Regulators (CEER), which reviews energy regulatory development in south-eastern Europe, a relatively undeveloped region (Vandenbergh, 2004). All the countries reviewed now have laws creating a regulatory authority as a separate legal entity. The majority of the authorities have budgets separate from the central budget and may issue decisions without the approval of a governmental body. They typically review tariffs and investment plans. Most have power to settle disputes, as well as some rule-making power. In addition, the majority have accounting systems for unbundled activities and procedures for some level of public participation in regulatory hearings. Staff size is mainly in the range of 15 to 60.

A few countries held out against a separate regulator, preferring to rely on their competition authority to manage issues of competition and otherwise letting the market take its own course. New Zealand started with a framework based on voluntary agreement between the players. This led to the development of a sophisticated market place, but further development stalled and a regulator (Electricity Commission) was established in 2003. In a controversial experiment, Germany went for full market opening without a regulator. Although prices fell immediately after market opening, subsequent industry mergers consolidated *de facto* monopolies, leading to a decision to establish a regulator.

The following list summarises the basic requirements for regulators – *i.e.* clear strategic objectives, a broad range of specific responsibilities and linked enforcement powers. This combination is needed to ensure that the electricity regulator can exert effective influence over the power market.

Strategic objectives

- Regulatory oversight of the relevant entities (TSO, ISO) for transmission and system operation.
- Development of a competitive market.
- Oversight of universal service and social obligations.
- Co-ordination with environmental policy.
- Promotion of energy efficiency.
- Advice on policy and further reforms.

Specific responsibilities

- Manage the regulated third-party access to the grid, normally associated with related issues such as the management of unbundling (absent full divestiture).
- Establish or approve the calculations for network revenue (some regulators are directly responsible for this function).
- Fulfil an oversight role in approving new investments, taking into account that transmission is a monopoly and the costs of new investments will be borne by all parties and ultimately by consumers.
- Grant licences to network operators and other market players.
- Control prices for transmission and for end-user electricity supply (if the latter is still regulated).
- Set rules for system operation and power markets.
- Approve grid investment plans.
- Conduct data collection and analysis on the power sector, thereby significantly increasing transparency.

Linked powers

- Powers of enforcement in respect of those issues for which it is responsible (including the possibility of inflicting direct penalties for non-compliance).

- Powers to require reporting of information from recalcitrant regulated parties (even though it is, for example, difficult to police grid access if the regulator suspects, but cannot prove, the existence of cross-subsidies by a dominant operator).
- Powers to resolve disputes in relation to issues such as grid access, grid use conditions, grid charges and end-use electricity tariffs for captive customers.
- Powers and capacities for data collection and analysis, so that the regulator has a good grasp of power sector developments (including projections of future demand and supply), and does not have to rely on the entities which it regulates for this information.

Essential features of an effective regulator

An essential institutional characteristic of an effective regulator is a clear and credible authority over the entities which it regulates. Hierarchy is important in China's governance system. Thus, the specific challenge in this context is to ensure that the regulator has adequate authority over the SOEs that it needs to regulate, including and not least the SG, which is one of the largest grid companies in the world. In China, SOE heads are appointed directly by the Communist Party and therefore hold a great deal of power.

Greater transparency for SERC's actions and advice is also essential. The aim should be to work towards a system under which SERC's proposals (including its recommendations to the NDRC and other policy makers) are systematically made public, with the opportunity for public comment on the proposals. Similarly, SERC should make a public accounting of its own actions. This openness would help SERC achieve the higher profile that it needs to develop.

At present, SERC is too small. An effective regulator must have adequate staff and resources. If staffing is low, it can undermine independence: the regulator then needs to rely too much on other government bodies for support. There is a relative absence of "regulatory economies of scale" in this sector. The costs of a well-staffed regulator will be low relative to the expected benefits of reform. Staff competences are also important, and should include industry and consumer experts, as well as specific competences in economics, law, and accounting. A very general issue for all new regulators is the need to develop an understanding of how companies behave in competitive power markets. Armed with this understanding, China's regulator would have a better chance of taking effective action to deal with abuses of market power. A training programme to develop staff may be useful, including, and perhaps especially, for staff in SERC's provincial branches. The power companies could be required to pay for the costs of SERC.

Beyond the transition: strengthening SERC's accountability

As SERC develops, there will be a need to establish adequate safeguards against misuse of regulatory authority. There are political risks for investors if they cannot be confident that regulators will be fair and impartial – just as there are political risks if the regulator is too close to policy and political decision making in government. A balance is needed in both directions in order to allow the regulator latitude to regulate effectively within a framework under which its decisions can be challenged, if necessary.

The starting point is basic transparency. Public information about activities needs to flow to stakeholders via Web site information, public hearings, annual reports, etc. Consultation prior to important decisions, and publicising those decisions, helps to avoid capture by special interests. Another useful approach is the establishment of clear strategic objectives against which performance can be measured (to develop a more competitive market, to protect the consumer, etc.). Regular auditing has proven useful; in fact, in many countries, the national audit office has proven to be an unexpected ally of reform. The annual reports of China's State Audit Administration evaluate the actions of the Ministry of Finance and other government bodies at the central and local levels, and could also be deployed in the power sector. Eventually SERC could be made accountable to the National People's Congress.

Strengthening regional and local capacities

Because of China's size, reform implementation is largely in the hands of sub-national levels of government. However, China's provinces are at very different levels of development, which affects their ability to manage reform. This is reflected in the decision to experiment with competitive power markets in three of the more advanced provinces before expanding markets further. Conducting the experiments in these provinces provides a test-bed and benchmark for emerging best practice. This should also be combined with policies that seek to strengthen the regulatory capacities of the local levels and that prevent the fragmentation of a coherent, centrally driven regulatory framework.

Two of the main challenges will be non-discriminatory system dispatch and grid access. Despite the unbundling, generators and the grid companies have retained close links with local government, which also has influence over some major users. The task of reducing the impact of this influence will lie with SERC's local branches. The weaknesses of these branches – lack of experience of independent regulation, a relative lack of influence, and the fact that many of the staff are drawn from the local area – will need to be addressed. To ensure that these local branches are not captured by either the local government or by the local power companies, the best approach might be to fund the local branches from the centre, and to ensure that staff is recruited from a nationwide pool. This principle should apply equally to grid companies. However, this runs counter to the tradition of local staffing of local agencies.

Recommendations for strengthening institutional capacities

Appoint SERC as independent regulator. China must take action to strengthen SERC's capacities so that it can evolve as an independent regulator. This includes ensuring that SERC is empowered to take the necessary regulatory actions in the market as it now stands, without waiting for further developments in competitive power markets. These actions should clarify SERC's objectives, responsibilities and powers – and strengthen its institutional features.

Clearly define SERC's mandate. SERC's objectives and responsibilities should be clarified and communicated to all stakeholders. In addition, SERC should be provided the enforcement powers needed to discharge these responsibilities effectively, across all of the following areas:

- Pricing.
- Regulatory oversight of market players (generators and the grid).
- Oversight of system dispatch and system security.
- Shaping the regulatory aspects of energy efficiency and environmental policy development.
- Data collection and analysis of the power sector.

Prepare SERC for enhanced role in pricing. SERC should gradually develop an enhanced role in pricing decisions, with the long-term goal of taking over this responsibility. In the near term, SERC should be formally empowered to make the advice it provides to NRDC easily available to broader publics. SERC should be given responsibility for monitoring costs and enforcing the pricing rules.

Empower SERC to address anti-competitive behaviour. Pending the establishment of a competition authority, SERC should develop its capacities for identifying and monitoring anti-competitive behaviour. Staff with the necessary competences should be recruited or trained, perhaps through exchanges with other jurisdictions that have successful track records in dealing with these issues.

Strengthen SERC's institutional features. In order to support the expansion of SERC's institutional role, it is critical to, in the first place, strengthen staff and resources. Effort should also be made to increase transparency, with the aim of creating a stronger public presence for SERC, both within the market and with the wider public.

Enhance regulatory capacity across government. Steps should be taken to enhance the regulatory capacities of local, as well as national, levels of government. As China has frequently done in other areas of reform, stronger provinces could be allowed to experiment with markets and market regulation, in order to provide test-beds and to benchmark emerging best practices. Actions to support local development might include the transfer of experienced officials to poor areas, flexible pay scales, and, for the more senior posts, continued input from the central government on recruitment.

PROMOTING ENVIRONMENTAL GOALS: TACKLING COAL POLLUTION[27]

The continuing importance of coal, and its environmental effects

Regardless of origin, all forecasts, scenarios and plans concerning power generation in China point to decades more of coal's dominance. A major energy strategy study released in 2004 is typical, finding that, under varying assumptions, coal may account for between 59% and 70% of generation capacity in 2020 (China Energy Development Strategy and Policy Research Group, 2004). For this reason, the power sector is and will remain the largest emitter, by far, of some of the most important

27. Specific recommendations related to this topic are given at the end of this section.

airborne pollutants in China. Emissions of sulphur dioxide are commensurate with its share of coal use, accounting for 44% of the national total in 2004, the latest year for which data are available (NBS, 2005). The next largest sector, residential and commercial buildings, emitted only about half that much. Because most power plants operate particulate emissions controls, the electricity sector share of particulate emissions is much lower, about 19% of the total (building materials is the largest at 37%, dominated by cement manufacturing). As sulphur dioxide (SO_2) and particulates are the two air pollutants of greatest concern to China, the power sector looms large in efforts to curb emissions increases.

While China is increasingly active in the area of climate change mitigation, this issue has not yet risen to prominence in domestic policy. China has, however, begun to track emissions of greenhouse gases to comply with its international reporting obligations. Shares of carbon dioxide emissions reflect fuel consumption even more closely than those of other pollutants. This is another area in which power plants are the largest source.

Stronger policies could have a major impact on coal use and emissions

An analysis of the model results for China prepared for the IEA's *World Energy Outlook* 2004 show that implementing stronger policies would significantly reduce growth in coal use and in the resultant carbon dioxide emissions in China's power sector. In Figure 7, the set of charts on the left show detailed results for the power sector from the "Reference scenario", which represents baseline assumptions about policies, technologies, and driver variables such as economic and population changes. Charts on the right show results from the "Alternative scenario", which depict the possible impacts of policies and measures many countries are considering adopting. In the case of China's power sector, the policies assumed to be carried out included: refurbishment of existing coal-fired power plants; expanded support for greater efficiency and cleaner coal-fired power plants; expanded government support for gas-fired power plants; extended support for generation based on renewables; policies to promote combined heat and power; and more government support for nuclear power.

Neither the Reference nor the Alternative scenarios are intended to be predictions; rather they show the possible effect under assumed conditions of various factors related to energy supply and use. With this in mind, it is still clear that there is considerable potential for decelerating growth in coal use and associated emissions. By 2030, the Alternative scenario shows a total installed capacity 150 GW (or 13%) less than the Reference scenario, mainly due to a 26% decrease in coal-fired capacity (although coal-fired capacity in the Alternative scenario still more than doubles between the base year and 2030). The shares for changes in power generation are similar. Because of the slower growth of coal capacity and its partial displacement by less carbon-intensive alternatives in the Alternative scenario, emissions of carbon dioxide are nearly 1 billion tons (25%) less than in the Reference scenario. Moreover, the trend in emissions in the Alternative scenario becomes one of visibly decelerating growth. Much of this observed change would be driven by policies on the demand side that would reduce growth in consumption of electricity in industry and buildings. This demonstrates how supply- and demand-oriented measures can complement each other in achieving environmental goals.

Figure 7 Reference and Alternative scenarios for coal use and CO_2 emissions in China's power sector

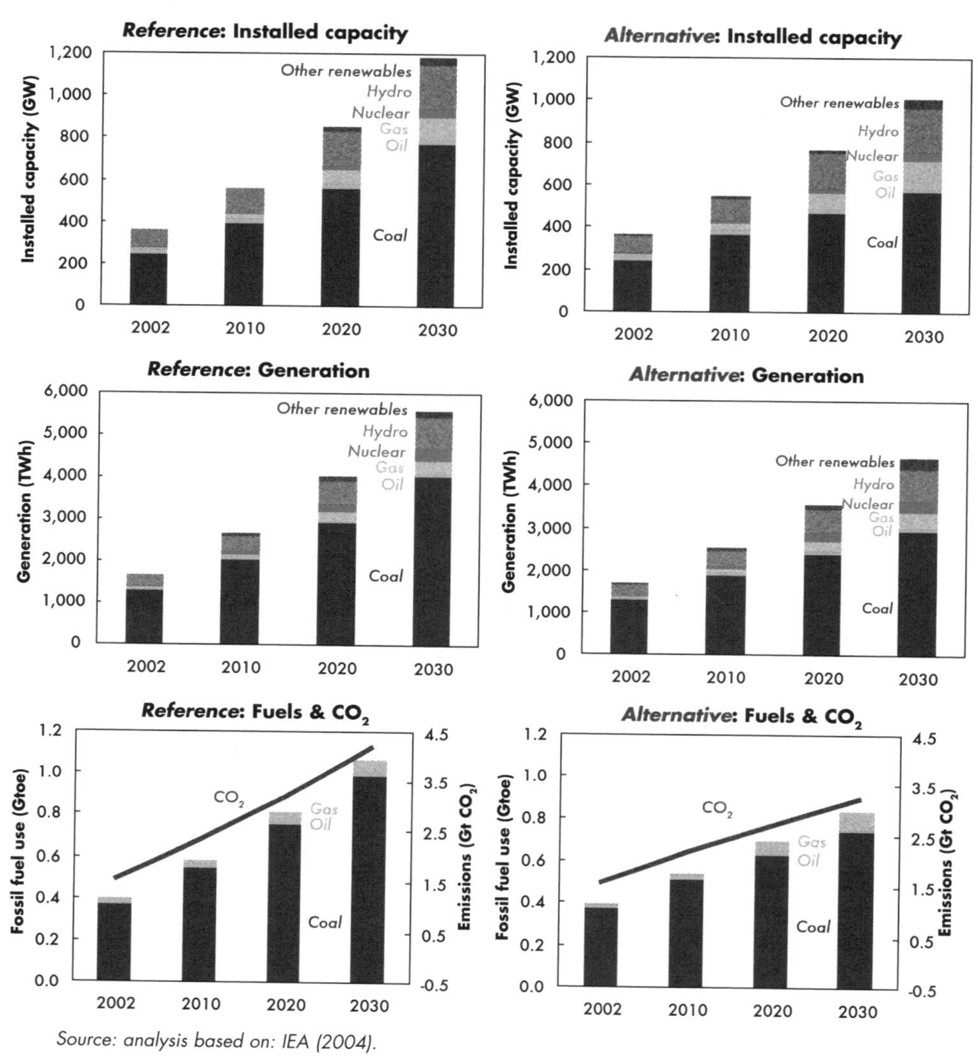

Source: analysis based on: IEA (2004).

The main sources of improvement in environmental performance Improvements in the environmental performance of power plants arise from three main sources: changes in fuel; improvements in efficiency; and emissions controls that either reduce the amount of pollutant precursors in the fuel (*e.g.* washing of coal) or take pollutants out of the waste stream (*e.g.* flue gas desulphurisation or FGD). These latter two categories both fit within the definition of clean coal technologies (CCTs), which comprise a spectrum of technologies, from mine to smokestack, that all contribute to reduced emissions.

Fuel switching Switching away from coal is highly problematic in a country where coal is abundant and relatively cheap. Even in coastal areas, where domestic coal prices are high enough to make imported coal worthwhile in some cases, coal is a fuel of choice for its ease of storage, ease of use, and, usually, easy availability. All alternatives present difficulties that prevent their rapid adoption on a scale that would significantly displace coal. Natural gas prices are high and the fuel goes preferentially to high-value uses in homes, commercial businesses, and industry. In addition, imported piped gas from Russia and Central Asia is some years away. Supporters of gas-fired power generation have been calling for preferential policies to be adopted by the central and local governments, as have supporters of non-hydro renewables. Oil is, like gas, relatively expensive and is especially valued for transportation and petrochemical uses. Nuclear power is very expensive and takes a long time to build. Hydropower is also slow to come on line, capital-intensive (for both generation and transmission assets), and undermined by environmental and social issues. Wind power is growing in importance, but will begin to encounter resistance of various sorts before it can grow to rival hydropower in scale. These other generation sources can contribute to reducing the emissions burden created by the power sector, but it is clear that major improvements are also needed in coal technologies.

Efficiency improvements

The efficiency of Chinese coal plants has improved steadily, but remains low in international comparison. Net generation efficiency was nearly 33% (excluding units under 6 MW, which are not included in statistics) in 2005, compared to 30% ten years ago (National Bureau of Statistics, 2001; Energy Conservation Information Dissemination Center, 2006). While China has made considerable progress, a typical average worldwide is about 37%, suggesting that China has about another decade of progress – and several hundred GW to install – before it reaches that mark. Further efficiency improvements have significant payoffs. At current rates of power generation, every 1% point of improvement in overall power generation efficiency results in coal savings of about 20 million tons per year, offsetting the need for roughly 11 GW of new capacity, and avoiding approximately 37 million tons of CO_2 emissions per year.

Much of the recent improvement is the result of new additions with unit capacities of 300 MW or higher. Despite prohibitions that the central government issues periodically, small, inefficient units are still being built, which bring down the average efficiency of the power sector. In 2004, 45% of the new capacity additions were in units of 300 MW or larger; nearly 10% (almost 4 GW) were in units of 24 MW or less. The bulk of these were likely to be diesels sets, but small coal-fired units continue to be built as well.

Emissions controls

Pulverised coal-fired power plants, whether sub-critical or super-critical, will be the mainstay of the sector for some time to come. This highlights the pressing need for improved emissions control, specifically capture of particulates and SO_2 emissions. Control of particulates is relatively easy and inexpensive, and is a well-accepted practice in China. Control of SO_2 has proven to be much more difficult. Essentially, it means installing – and running – FGD equipment, which is expensive, uses a great deal of electricity that a power plant could otherwise sell, and produces an additional stream of solid waste. For more than ten years, China has been subjecting a larger number of power plant projects to the requirement to install FGD, or at least to design plants such that they can be easily retrofitted for FGD. However, this initiative has had very limited success. Currently, only a few dozen power plants in China are fitted with sulphur scrubbers.

Sometime in the future, all coal-based plants, including integrated gasification and combined cycle generation (IGCC) plants and facilities that produce an array of coal-based energy and chemical products, will require carbon capture and storage technologies to control carbon dioxide emissions. That time is yet far off, and experience in promoting FGD deployment should provide some valuable lessons.

Implementation and enforcement of environmental regulations

How can these potential changes be realised? The scenarios for lower pollution are not dependent on rapid deployment of exotic new power plants. They imply, instead, a more vigorous deployment of demand-side policies, together with the adoption of off-the-shelf technologies that, in many cases (like supercritical coal-fired power plants) are significantly better than China's typical new plants. China already has experience with these technologies and is already deeply engaged in the worldwide effort to develop and apply CCTs (Philibert and Podkanski, 2005). The country is unlikely to delay long in adopting proven technologies that are successfully applied in other countries, provided that China has in place an appropriate regulatory environment and set of incentives – including the means to strengthen the system of environmental protection. This implies integrating environmental issues very deeply into electricity policy and regulation.

One should not, of course, dismiss the important technical and financial issues associated with the adoption of more efficient CCTs. For instance, IGCC is one much-promoted technology that would bring great efficiency gains and emissions reductions. However, IGCC has had great difficulty gaining a commercial foothold – even in developed countries – because of costs, technical challenges, and perceived risks. The challenges involved in persuading power sector players to choose and use less-polluting coal technologies, as well as alternative generation sources, fall within the supply side of the reform process. They include: strategic integration of environmental goals into economic regulation of the power sector, institutional structures that bring the two together, pricing, and market regulation. These are reviewed in more detail in the section on the regulatory framework.

To a large extent, the regulatory and institutional reforms identified in other sections of this report, along with strengthening SEPA's enforcement powers, are the keys to implementing and enforcing environmental regulations. Like most countries, China has long struggled with the issue of enforcement of environmental statutes. This report is not the place to address those issues in depth. Following the theme of transparency raised throughout this report, one way forward would be to use pressure from public opinion to complement enforcement efforts. SEPA has become interested in public disclosure because China's pollution problem remains severe, despite long-standing attempts to improve environmental quality with traditional regulatory instruments. Informal and *ad hoc* channels of canvassing public views have been employed in many situations. The time is now ripe for formalised means of engaging citizens via public hearings, advisory committees, information meetings, and reviews of environmental impact assessments and other documents. Such activities would help to diffuse social tensions while ensuring adequate consideration of a broad cross-section of views.

Integrating environmental issues into regulation and competition

The ongoing reforms and development of a new regulatory framework to introduce competition provide an opportunity to consider the impact on environmental protection. Indeed this should be considered a necessity. Absent regulation that incorporates incentives to adopt more environmentally friendly options, the broad effect of competition – and of a regulatory framework that focuses solely on competition – is likely to produce negative environmental effects. If there is to be cleaner power, these effects must be countered via a review and adjustment of the regulatory framework. Two preferred options would avoid distortions to market decisions:

- **Pricing and plant dispatch.** Incorporating environmental costs and benefits in power pricing, including at the generation level, could be expected to have significant effects on investment decisions and plant dispatch. Plants with low operating costs will run more and those with high operating costs will run less. For example, coal plants without FGD will run more and those with FGD will run less. Natural gas fuelled plants have higher operating costs and may also be expected to run less. This is entirely consistent with the principle of cost efficiency, and competitive markets are designed to encourage this. Environmental considerations therefore need to be factored in, with care. Establishing uniform generation performance standards or a pollution fee would put the responsibility on the market to work out the most economically efficient mix of FGD and other pollution control options to meet the requirement.

- **Investment planning methodologies and licensing rules.** These will affect the type of plant that is built. Unless the rules lend support to other technologies, coal may be favoured. Environmental costs and benefits need to be appropriately reflected in generation and grid investment decisions. Licensing requirements also need to be strictly enforced to ensure that unauthorised (and likely dirtier) plants are no longer built[28]. Including environmental costs and benefits in the planning process for new power plant investment would encourage investment in cleaner plants. Indeed, it could be argued that China is implicitly adopting this approach already, at least to some extent, in the investments that it has made in nuclear and hydropower plants. But there is scope for a more consistent and rigorous review of investment mechanisms to ensure the capture of environmental costs and benefits of various options.

The regulatory framework may also be adapted to assist in the enforcement of environmental regulations, which is a known weak spot in China. In particular, two types of support for environmental levies could be implemented:

- **Through market design.** It will be difficult, if not impossible, to implement environmental policies to reduce emissions without a framework that enables the tracking of emissions. The design of new competitive markets should seek to incorporate emissions tracking systems. These can help to monitor environmental effects, support enforcement, and help with the collection of environmental taxes or levies. In the longer term, with the advent of retail competition, these systems will enable consumers to choose between different types of power depending on their "greenness".

28. A related issue which is outside the scope of this report is the exploitation of unauthorised coal mines.

- **Via the grid companies.** As long as the grid companies retain a central role in power sale transactions (as they do now as single buyers), they are well positioned to collect environmental levies.

Taking these ideas forward requires a robust and adapted institutional structure, capable of ensuring the development and implementation of sound policies that strike the best balance between competition and the environment. This implies reviewing how the NDRC and SERC can best link into the work of the environmental agencies. SERC could be tasked with the specific responsibility (alongside its other responsibilities) of integrating environmental goals into the economic regulatory framework. It would need adequate staff resources and competences for this. An institutionalised co-operation mechanism between SERC and SEPA would also be valuable. It could, for example, provide for regular meetings between the two agencies, and staff exchanges. Likewise, SEPA might be specifically tasked to assess the environmental consequences of reform in proposals for the power sector.

Measures to promote and integrate demand policies and energy efficiency, which are considered elsewhere in this report, are also highly relevant to the goal of cleaner power. They reduce the amount of power that needs to be produced and, hence, the amount of pollutants derived from power production. The *IEA World Energy Outlook 2004* demonstrates potential effects with its Alternative scenario, which relies heavily on the implementation of demand-side policies.

Recommendations for promoting environmental goals and tackling coal pollution

Integrate environmental goals in policy reform. China should review and adjust its developing regulatory framework for competitive markets to ensure that it supports environmental goals. Areas that require particular attention include:

- Establishing fees or emissions standards to help secure the dispatch of cleaner plants.
- Incorporating environmental costs and benefits in power pricing.
- Conducting a review of investment planning methodologies and licensing rules to encourage cleaner investments.

Create mechanisms to enforce environmental regulation. China should seek to adapt its regulatory framework to support the enforcement of environmental regulations. This might include actions that help to track emissions and involving grid companies in enforcement activities.

Link environmental goals with market competition. China should review its institutional structures to ensure that they are capable of promoting the development of policies that ensure consideration of environmental goals as competition develops. NDRC and SERC should establish means of linking into the work of the environmental agencies. In addition, an institutionalised co-operation mechanism should be established between SERC and SEPA to take advantage of the complementary nature of these two agencies. SERC could be tasked with the specific responsibility of integrating environmental goals into the economic regulatory framework; SEPA could assess the environmental consequences of reform proposals for the power sector. Both agencies would need adequate staff resources and competences to do this; a formal co-operation agreement could create opportunities to, for example, provide for regular meetings and staff exchanges between the two agencies.

TOWARDS MORE EFFICIENT PRICING AND INVESTMENT[29]

More efficient and cost-reflective pricing will make a core contribution to China's strategic goals for the power sector. In fact, getting pricing right will go a long way towards improving both the economic and energy efficiency of China's power sector. The case for the further development of cost-reflective pricing for China's power sector is compelling and increasingly urgent. Despite periods of adequate supply, power shortages recur with uncomfortable regularity as the sector struggles to keep up with rapid economic growth – which is expected to continue into the foreseeable future[30]. The shortages, when they occur, affect economic performance (factories close down to save energy) and undermine the goal of adequately providing – at all times – for the essential energy needs of a society. Power surpluses are also damaging for efficiency. The most important, and fundamental, argument for additional reform is that more cost-reflective prices across the value chain would provide signals to trigger efficient investment and to curb consumption. Along with other incentives for demand management, and better investment planning, a more efficient pricing framework is a key mechanism for meeting the energy goal – both now and over the long run.

A second argument for the development of cost-based pricing is the opportunity cost to the economy of subsidised power and of inefficient investment, which is still largely by the state[31]. In particular, there is growing indebtedness to the state-owned banks. Outside of share issues on overseas stock exchanges by several of the largest generating companies, investment from the private sector, both foreign and domestic, remains small. In other words, the power sector may be draining resources away from public policy priorities such as investment in health and education, and support for migrants to urban areas. A key goal of further reform should be to put the power sector on a footing where it can sustain itself, so as to reduce and eventually eliminate public funding.

A third argument for cost-reflective pricing is that it will likely encourage consumers to consider their energy consumption and ways to reduce it.

The reforms to date have been mainly successful in encouraging investment, but rather less in encouraging efficient investment, which promotes demand as well

29. Specific recommendations related to this topic are given at the end of this section.
30. Optimism that the problem has been fixed at last is also a recurring theme. For example, the NDRC is quoted as saying that "The year 2006 will see the end of electricity supply shortages" (China Daily, February 20, 2006). The main arguments in support of the statement are that generating capacity has increased and energy-saving measures are bearing fruit. But the history of recent years shows that despite improvements in generating capacity and energy efficiency, power shortages continue to emerge regularly.
31. The OECD's Public Governance report (2005) notes that there are three sources of pressure on public finances: restructuring of the state-owned sector, the risk of bank insolvencies, and rising demand for social welfare. The less efficient SOEs, therefore, face increasing financial difficulties as they can no longer be subsidised, and profitable companies are milked.

as supply-side investment and efficient behaviour by generators. Incentives designed to lower costs and increase productivity remain very weak, as do incentives to invest in end-use energy efficiency. A key issue is that the current framework does not encourage investment in end-use energy efficiency as an alternative to supply-side investments. A main goal of generators, which remain largely publicly owned, is to increase market share and make profits (which may then be used to promote local social and economic development). Some generators could be making super-normal profits under the current framework, *i.e.* profits above those that might be expected in a well-functioning competitive market. This amounts to a misallocation of resources. At the same time, the current framework does not allow effective pass-through of generators' costs to end users; these costs are increasingly vulnerable to rising coal prices, which are set broadly by the market. Some generators are increasingly squeezed by rising coal prices, but are blocked from passing through the extra costs to consumers. This dampens the incentive to invest.

As regards the grid, the absence of separate grid pricing, particularly in a system of non-cost-based grid planning, creates a situation in which grid investment lacks the efficiency signals that a clear cost-reflective planning and pricing methodology would provide. This reduces the likelihood that such investment will be optimal.

In reality, reforms have not yet established a framework for fully cost-reflective and efficient pricing across the value chain, even though there have been some important changes over the last decade or so. The price of coal – the most important fuel input to power generation – has been largely liberalised. Generation pricing broadly seeks to reflect costs, though very imperfectly. However, efficiency is not encouraged under the current pricing system. End-user pricing is now generally close to marginal cost, though again, there is no clear methodology to underpin this cost-reflectiveness and to ensure that it continues into the future. Decisions about end-user pricing levels tend to be more politically based. The regulation of end-user prices is used to meet social and economic objectives (customer affordability, the promotion of a stable economy through stable prices, and protecting the competitiveness of downstream industry) rather than to promote efficiency in the power sector itself. Grid pricing does not yet exist, nor is there any distinction between wholesale power tariffs and tariffs for grid use (one tariff covers both). Power tariffs are currently very complex: over time of many different calculation methodologies and charges (for example, charges imposed since the late 1980s to collect funds for developing selected generation and grid projects). The central government (NDRC) retains a tight control of power pricing.

The State Council's 2003 policy document *Scheme for Power Price Reform* tabled proposals for further reform linked to the development of competitive regional power markets. A follow-up document in March 2005 set out the plans in more detail. The proposals have some positive features, notably the establishment of separate grid tariffs. Wholesale generator tariffs would be in two parts: a capacity component (set by the NDRC) to encourage investment in new plants, and an energy price, set by competition in the new regional power markets. Only modest changes are proposed for end-user pricing[32]. Residential users will continue to be subsidised by other users, and there is still no clear guiding principle or methodology that would ensure cost-reflective pricing for this category. In other words, generators (and grid managers) cannot (as now) expect that their costs will be passed through to this group. This weakens the incentive for generators to invest if they believe that they may not be able to recover their costs. Ultimately it is likely to undermine needed investment in the grid.

The institutional arrangements for price setting raise a major issue. As now, the government would remain in charge through the NDRC pricing department (and its provincial pricing bureaus), not the regulator SERC. With the political arm of government responsible, rather than the regulator, pricing decisions are likely to be influenced more by political considerations than by what is best for a well-functioning competitive market.

Implementation delays – it is three years since the first pricing reform announcement – suggest that the changes proposed, limited as they are in terms of developing a complete cost-reflective approach, still raise political difficulties. Such delays are very damaging for the credibility of the government's commitment to developing reforms that will eventually lead to competitive markets.

There are four main issues with the current framework, which proposed reforms address only partially – in the case of energy efficiency – or not at all – in the case of the environment. Addressing the issues requires a large effort to increase transparency along with other actions to reform pricing for generation, the grid and end users. It does not require the implementation of fully competitive markets. In fact, pricing approaches will need to be adjusted again as markets move towards full competition:

- A more transparent approach to pricing, linked to the application of cost-reflective methodologies, would help to identify the extent of use of public funds in the power sector. It would also start to shift the power sector away from dependence on these funds, towards a system that pays it own way.

- Cost-reflective pricing is not yet applied across the whole value chain and there is not, as yet, separate pricing for each service component. System dispatch, based on current pricing, is not efficient.

- Pricing does not include incentives for demand-side investment and end-use energy efficiency, nor for least environmentally damaging options.

- Grid investment planning is not addressed in the most efficient way.

At this stage, the overall objective should be to ensure that prices throughout the value chain are based, as far as possible, on costs – including incentives and provisions

32. There is no problem as such with the continued regulation of end user prices, in the absence of full consumer choice of supplier. Most countries continue regulating end-user prices until this point has been reached. The issue is rather the methodology for regulating prices, which does not seek to ensure that prices reflect costs.

for energy efficiency and environmental costs. Absent competition in the near term, this means putting in place a system of regulated prices that does this while ensuring that costs can be passed through and that energy efficiency and clean investment incentives are built in.

Four near-term actions provide first steps towards more efficient and cost-reflective pricing

Given the complexity of current pricing systems and the need for regulator reform in other areas, truly efficient and cost-reflective pricing is still some time away. However, important steps towards this end goal can be undertaken in the near term, particularly in relation to wholesale pricing and dispatch, transmission pricing and grid dispatch, grid investment, end-user pricing, and input fuel (especially coal) pricing.

Wholesale pricing and system dispatch

Wholesale prices have been allowed to rise to better reflect capital and input fuel costs. However, the pricing methodology applied across most of the country does seek to promote cost reflectiveness in a broad way. Ergo, this results in inefficient plant dispatch and does not promote generation efficiency. Most of the country (the exception is the Northeast regional market) currently applies a single, technology-differentiated contract price for generation. The order of plant dispatch is based on a combination of this price and a prearranged number of hours that the plant is expected to run over the year (both elements form part of the contract – or power purchase agreement (PPA) – that generators have with the grid companies). This form of dispatch does not reward low marginal costs, which tend to be characteristic of the more advanced plant types. Advanced, clean and efficient plants are paid more, but they are dispatched less often; hence, they operate fewer hours than less efficient plants. In short, the current pricing and dispatch arrangements do not reward efficiency.

China proposes to implement a two-part generation price. This would comprise a capacity price that reflects the plant's capacity (or capital) cost, and an energy price that reflects the plant's energy (variable or marginal) cost. The capacity component would be paid for on a per kWh per period basis, whether or not the plant produces power (though there could be penalties for non-performance). Power would be paid for according to the number of kWh actually produced, set to cover the marginal cost of the plant's operation. The power price would then be the price on which dispatch decisions are made (cheapest going first, etc.). This approach helpfully promotes more efficient plant dispatch, based on each plant's marginal cost. When fully competitive markets are established, the capacity component should be reviewed. Capacity pricing in fully competitive markets is controversial, as it masks pricing signals for efficient investment.

Transmission pricing and grid investment

China does not, as yet, have any separate pricing for the grid, or even separate accounting for grid costs. This not only distorts the picture for generation pricing, but also means that there is no clear framework for determining efficient and adequate grid investments. Moreover, there is no mechanism for players in emerging regional power markets to establish efficient trade, based on the cost of power and the separate cost of using the grid to dispatch the power. Grid revenues currently derive from the difference between the revenues that grid companies earn as single buyers of power from the generators, and the prices they charge for selling the power to retail customers, topped up with special funds for specific investments. Thus, the prices set for generation implicitly include an element to cover grid transportation. However, this element does not derive from any methodology to establish the true cost of transportation.

China proposes an important reform to establish separate grid pricing, based on the postage stamp approach for user charges[33], and a cost-plus approach (a form of rate-of-return regulation) for grid revenues. This combined approach would be used to set transmission tariffs across each regional market, and to set distribution tariffs at provincial level.

The strategy for user charges has one major advantage: its simplicity. Other tariff systems that take into account line losses and transmission constraints would be highly complex, requiring strong technical and economic analytical capacities and effective data collection. The risk of regulatory failure needs to be taken into account with more complex systems, and China should be careful to ensure that costs do not outweigh benefits. The postage stamp approach also has the merit of avoiding pancaking – *i.e.* adding tariffs at jurisdictional borders – which is a huge barrier to trade. Simplicity is perhaps the main advantage of a postage stamp approach; it also has a number of serious drawbacks:

- **It does not promote economic efficiency and investment.** The postage-stamp approach is inefficient because it does not reflect true costs. For example, a company deciding where to locate a new factory will not consider the higher cost to the electricity system of locating that factory a long distance away from generation facilities. Incentives to invest, and to do so efficiently, are very important for such a large country with a need to expand and develop its grid. The priority is a tariff system that encourages efficient investment in new transmission infrastructure and sends appropriate locational signals for the efficient siting of new generators. The implications of a weak grid and interconnections are important for the prospects of developing competition and trade across China's power markets. There is the risk of reinforcing regional monopolies if power cannot "travel". Without fully cost-reflective pricing, there is a risk that the revenues earned by the grid companies will be inadequate to finance the much-needed expansion of the network. If reforms fail to ensure adequate investment, China runs the risk of falling back on a wholly planned approach to investment, not just for the grid but generation as well. The cost-plus approach to grid revenues is also less than optimal, as it does not provide any incentives for the grid owners to invest efficiently.

- **It does not promote energy efficiency.** Grid companies are an important target for the promotion of energy efficiency through demand-side investments. The methodology for establishing their revenues is critical to providing incentives that encourage them to consider the demand side. Before generation was separated from the grid, China's original integrated utility had some incentive to invest in or

33. This is the traditional approach. All customers pay the same rate, regardless of differences in cost related to location. Costs or permitted revenues of the transmission system are instead shared among grid users depending on their generating capacity, peak demand, total generation or total consumption.

encourage energy efficiency, for example during the periods when generation costs exceeded retail revenues from additional sales (*i.e.* when marginal cost was greater than marginal revenue). These incentives no longer exist. Grid companies need financial encouragement to promote energy efficiency. It is important therefore, as part of an overall strategy, to promote end-use energy efficiency and to consider how regulation of grid revenues can encourage the grid companies to take account of the demand side when making decisions about investment in grid infrastructure. Revenue-based regulation can, for example, create incentives for grid companies to reduce congestion costs and encourage them to look at energy-efficiency measures as way of dealing with this issue. It is absolutely essential to avoid any type of regulation of grid revenues that makes it impossible for grid companies to choose to invest in energy-efficiency measures, because they will lose money this way.

- **It does not encourage balanced development and new renewables.** Any reform to the pricing approach needs to take account of China's objective of balanced regional development, in particular the economic development of Western China. The development of new renewables (wind, solar, etc.), which are often found in remote locations and cannot be "moved", is a related issue.

The postage-stamp approach should therefore be treated as an interim measure. Simplicity remains desirable until regulatory capacities have been strengthened. From the point of view of economic efficiency, nodal pricing comes closest to cost reflectiveness, but it is complex. Also, the experience of nodal pricing in other jurisdictions raises some issues, including its distributional effects. As discussed, there are other objectives for grid pricing. Developing the right approach also needs to take account of China's objectives of balanced regional development and the promotion of new renewables, and not least, rewarding investment in energy-efficiency measures.

Thus, postage-stamp pricing must be complemented with other measures that will introduce locational signals for the siting of new generation investments. Investment planning continues to be necessary, but it needs to be brought up to date to reflect the best practice of cost-reflective methodologies applied elsewhere. As well, if locational signals cannot be sent via grid pricing, they can be activated, to some extent, via the wholesale pricing of power (*e.g.* through auctions of cross-jurisdictional transmission capacity.

Grid investment

There is also an urgent need to improve the system for attributing revenues from use of the grid and grid-investment planning, and to ensure that funds are available – and are efficiently deployed – both for strengthening the grid and for investing in energy-efficiency measures. Because competition is not yet developed, China's current approach to grid-investment planning is not yet based on market signals. It is also very different from, and less effective than, traditional utility planning as it is understood and applied in most other countries. Traditional utility planning is usually based on models that aim to achieve a least-cost mix of different types of plant to meet the load curve. More advanced types of planning use similar techniques to integrate energy efficiency and environmental considerations into the planning process. Once the general shape of investment needs is determined, competition may be used to put in place the needed investment, in the most cost-effective way.

China's approach has been quite different. Strengthening the grid and expanding plant capacity remain linked to the traditional planning system and the five-year plans. It starts with a "bottom up" aggregation of data requested from the provinces and sub-levels of government. This is translated by the NDRC into an overall plan that sets an aggregate of targets and objectives for investment, production and consumption. This plan sets the course for the next five years and is repeated across a range of formal and informal documents from officials at different levels. Each individual investment must be approved by the government. Important decisions are currently being made under this system, notably the decision to go ahead with a major investment in the grid to strengthen east-west links across the country and to prioritise the national system over provincial systems. But are these the right decisions? A stronger grid is important for reliability and the prospects of developing competition in China's power markets. The grid also must be robust if it is to cope with a growing share of often intermittent renewable sources of power. But there should also be investment in energy-efficiency measures that would remove the need for some of these supply-side investments.

There is a need therefore, to think more broadly. Proposals to introduce a cost-plus approach (a form of rate-of-return regulation) for grid revenues should be aligned with efforts to develop a grid-planning process based on an evaluation of costs, as well as a mechanism that seeks to achieve the right balance between supply- and demand-side investments. The process for grid planning and investment should be progressively strengthened, with a view to improving its transparency and independence, and to ensuring that investment decisions are based on an assessment that takes account of the costs and benefits of various options. Encouraging further cross-regional trade and the gradual development of wholesale markets will help to identify bottlenecks and advance the process towards one that uses market signals to help guide the most efficient choices.

End-user pricing

China's end-user pricing regulation has evolved over recent years to include some very positive features. Pricing structures have been simplified by reducing the number of price categories and abolishing various surcharges, fees and taxes. There has been extensive implementation of time-of-use (TOU) pricing, as part of successful efforts to shift load in response to supply shortages. The differential between peak and off peak has increased. There has also been an increase in interruptible tariffs. Average prices now generally reflect marginal costs, though not for all end-user categories. Although prices have been allowed to rise, residential consumers continue to enjoy subsidised prices, as does heavy industry and agriculture. This disadvantages the commercial sector, which would include many small and medium enterprises (SMEs). SMEs will be vital to China's continued growth and modernisation, and a modern, efficient SME sector can contribute greatly to national energy efficiency.

At the same time, end-user pricing remains highly managed by the centre, without any clearly articulated principles to guide adjustments to the framework that determines specific prices (for example, to the time-of-day differentials or the addition of charges) or to changes in price levels. Pricing decisions by the NDRC are still influenced by a variety of factors, such as the need for social and urban/rural equity, without any strong and transparent grounding in a methodology that make cost reflectiveness a priority.

This contrasts with liberalised markets elsewhere, where end-user prices are either not regulated at all or are regulated within a framework that promotes cost reflectiveness and incentives to greater efficiency (revenue or price cap regulation)[34].

In addition to looking to develop and improve methodologies for more cost-reflective pricing – particularly for consumer categories that are currently under priced – another approach is to deploy a range of incentive and penalty schemes that encourage consumers to improve their energy efficiency. This approach has the advantage that it can be deployed quickly, whereas the unwinding of subsidies implied by cost reflectiveness is likely to take time because of the sensitivities involved. Both approaches are, in fact, needed over the longer term. China might consider several options in order to encourage energy efficiency among residential and other small consumers. Prices that focus on marginal consumption, or so-called inclining block prices, are structured such that the prices for blocks of consumption change incrementally as the consumer passes pre-defined usage thresholds. Another approach might involve linking prices to efficiency standards for buildings, via so-called "hook-up fees". In this case, consumers whose buildings meet very high efficiency standards would pay a lower price than those whose buildings are inefficient. The inclining block option has the additional advantage of keeping prices low for low volume, poorer consumers.

Input fuel pricing: coal prices for power generation

Cost-reflective pricing has advanced furthest in this part of the value chain, although some regulation still remains. The reform process has highlighted some of the difficulties of not implementing cost reflectiveness to the same extent in the other parts of the value chain. The liberalisation of coal prices, which are now largely set by the market, has led to a squeeze on generators' profits: the problem comes from rising prices that they cannot pass on to their own clients under the current pricing regime. It is not clear how far this squeeze affects their basic profitability. Profit margins may be high for some generators, in which case there is scope to digest the rise. Absent other cost-reflective pricing changes, coal prices have also contributed to the problem of supply shortage. Wholesale coal prices rose 40% in 2004, drawn upwards by world market prices. However, prices to power generators were lower and some generators faced coal shortages that reduced electricity production.

Another issue is that the coal price adjustment mechanism distorts investment decisions: it makes coal-fired generation less risky than other options. This removes the incentive for generators to invest in resources that may be less costly but higher risk, or from investing in hydro or other renewables, which are excellent hedges against fuel price risk. The mechanism also reduces the incentive for a generator to efficiently manage fuel costs, such as by seeking lower cost supply sources or hedging fuel price risk through longer-term contracts or other financial instruments. Finally, planned retail price reforms include the addition of a corresponding mechanism to adjust retail prices for changes in generation prices. This would distort the grid company's decisions to meet demand growth with power purchases, rather than investing in improvements to support end-use energy efficiency.

34. However, China's approach is common in developing and transition economies, in which the starting point is very different from developed market economies, and may be viewed as a natural staging post on the path to reform.

The political challenge: how to unwind subsidies without triggering social discontent

One of the big questions linked to reform is whether prices will rise if cost reflectiveness is developed further. It is a question that has many political implications, particularly in China where the power sector is a vehicle for other objectives: helping the poor; containing inflation; maintaining the competitiveness of downstream industry; and ensuring employment. Ultimately, these issues test the government's commitment to a market-based economy. For example, instead of using power sector prices as a tool for combating the economy's inflationary tendencies, greater flexibility in the exchange rate would be a more appropriate strategy (OECD, 2005b). The countervailing advantages to the economy of a power sector that is efficient and self-sustaining are considerable.

It is difficult to be precise about likely effects of reform on prices, as so many factors will come into play. These include (in no particular order): the approach taken to the management of stranded costs; coal and other fuel input prices; pricing in the cost of the environment; and the roll out of new, more efficient technologies. A key factor is the scope for improving generator efficiency. Long-term efficiency gains can be expected, which would reduce costs and contain prices. The realistic objective should be to constrain increases in retail tariffs rather than drive them down, which is likely to be unsustainable. Prices may rise for some categories of consumer (in which case, lifeline support can be set in place), but may decrease in other areas, thereby providing advantages to other important categories such as SMEs.

Social discontent can be a major issue with power market reforms. It is, therefore, important to balance the move to cost-based pricing with policies that address the social aspects[35]. Reforms generally encounter stiff opposition if they ignore this constraint. The relative inelasticity of power demand for households has important distributional implications. Subsidies are politically attractive because they are, in effect, the equivalent of a direct grant to households. Conversely their removal – by raising the price of power – has the effect of a direct tax that is not income-sensitive, and which therefore bears especially heavily on the poor and the elderly. In middle-income countries such as China, reforms to establish cost based-pricing and eliminate cross-subsidies can hurt poor people as well the better off.

Recent empirical work reported by the World Bank highlights the broad benefits of market-based reforms in the infrastructure sectors, which are positive for economic growth and so enhance economic opportunities for the poor (World Bank, 2004). When these effects are taken into account, the poorest groups seem to benefit the most from the increased productivity and access brought by reform. A study of Argentinean reforms found that all income classes benefited from the efficiency, quality and access improvements resulting from utility reforms. Other research on a range of Latin American countries (Argentina, Bolivia, Mexico, and Nicaragua) notes relatively small adverse effects on the bottom half of income distribution due to job cuts. However, this was more than offset by improvements in service quality, increased access for poor people, and changed structure of public finances, which had greater benefit for (all) poor people.

35. The World Bank (2004) has noted that this is a neglected area of research despite mounting evidence across a wide range of developing countries that dealing with the social factor in reform is a major challenge.

How do the poorer parts of the population fare under current arrangements? How much do subsidies actually cost? China's subsidies for power are opaque and there is no clear data on how much they might actually cost. Better understanding of the problem implies having a better grasp of how much subsidy is being paid, and who benefits. It also paves the way for sounder political decisions to address problems in the picture that emerges. It may become obvious that reform will need to be accompanied by "lifeline support" for those who need it. In this eventuality, two issues should be distinguished:

- **Protection of the most vulnerable members of society** (the poor and elderly). Even in wealthy countries, there will always be a section of society that struggles to pay its bills.

- **The rights of all household consumers to have access to reliable, good quality power supplies.** In developing countries, there is also the more basic issue of ensuring that everyone is connected to the grid, or at least has access to a stable source of power (through distributed generation, etc.). This is not yet the case in China.

China's vast and varied population encompasses a wide range of income levels; the blanket subsidisation of residential consumers does not distinguish between truly poor and more affluent consumers – and is not justified. A lifeline tariff is required for low-income families (urban and rural), but continuing an indiscriminate subsidy for an increasingly affluent population is unnecessary and sends the wrong signals for energy conservation and investment. There is also the current perversity that rural consumers often suffer add-on fees. Instead, vulnerable consumers should be explicitly targeted, and offered a lifeline. More targeted support for the truly vulnerable consumers may help to release funds for extending the grid in rural areas, in line with the government's desire to promote a more just and balanced society. It might also act as a political counterweight to the disaffection of other consumers, who may complain – in the short term – about price rises.

If lifelines are needed, many questions arise regarding how they should be designed and funded. The World Bank (2004) has proposed the following criteria for designing subsidies:

- **Effective targeting.** Benefits should accrue to the intended beneficiaries, such as poor people or rural populations.

- **Positive net benefits.** Subsidies should pass a cost-benefit test.

- **Administrative simplicity.** Schemes should have reasonable administrative costs.

- **Transparency**. Financial costs and payment channels should be clearly defined and open to public scrutiny.

One of the main questions is: Should support be funded from outside the power sector, or from revenues within the sector? In principle, funding should be part of general social support schemes: the power sector should not be used as a policy vehicle for managing broad development needs. In practice, the efficiency of the general tax system needs to be taken into account. A system that provides direct subsidies from power sector funds (*e.g.*, via a so-called "system benefit charge", which is a flat charge on the power price paid by all users) in a transparent way (the charge is identified in electricity bills[36]) may be the most practical approach. It does have the disadvantage of raising prices, though the increase can be expected to be small.

Targeting based on location or housing characteristics can substantially reduce subsidy leakage, and thereby significantly increase the share of subsidy resources captured by poor households. Targeted connection subsidies appear to perform better than targeted consumption subsidies. Yet subsidies have often been poorly targeted and failed to reach poor consumers[37].

The experience of the transition economies suggests a healthy degree of caution on the speed of implementing price adjustments. A transition period is needed, so China should start unwinding subsidies now. In some of these countries, the transformation to market economies has often been quite painful, particularly because of the contraction in economic activity and reduction in incomes. Moving too quickly to realign energy prices with costs can cause unnecessary social hardship[38].

In addition, it is vital to secure and maintain the right of all household consumers to reliable power. Thus, it is important to control, maintain and improve standards of service as markets open up to competition and new suppliers emerge. Residential users should have a right, regardless of their supplier, to connection and supply, and to a certain quality of power. Appropriate obligations should be considered for the relevant players. Obligations to meet quality of service standards may include the stability of power supply (voltage and frequency), the time taken to restore supply, and notification of interruptions. Licences may be used to spell out obligations such as service standards. This also aids transparency.

Recommendations for moving towards more efficient pricing and investment

Create transparent pricing approach that reflects real costs. China needs to develop an overall approach to pricing across the entire value chain that is transparent and reflects the costs of electricity production and transportation to end consumers. Ultimately, this will create a power sector that pays its own way and no longer depends on public funding, which could be better deployed elsewhere. This recommendation is linked to others, notably the need to corporatise the generation sector into well-defined companies with clear ownership, responsibilities and objectives.

Implement two-part pricing. China should implement its proposed two-part pricing principle to provide the basis for more efficient plant dispatch, based on each plant's marginal cost (*i.e.* system dispatch should be based on the power price, with the cheapest plant being dispatched first).

36. For example, the French gas/electricity utility EDF/GDF identifies the "contribution to the electricity public service" in its itemised bills to households.
37. In India, for example, state subsidies for water services totalled more than USD 1 billion per year (0.5% GDP) in the late 1990s – but poor households captured only a quarter of these (Foster, Pattabyak and Prokopy, 2003, quoted in World Bank, 2004).
38. In Ukraine for example, electricity prices were nearly six times higher in 1998 than in 1992, yet in the same period average household incomes fell by more than half.

Establish separate grid pricing. China should implement its proposal to establish separate pricing for the grid. However, it should also aim to move away from postage-stamp pricing in due course. The longer term goal should be to migrate towards a transmission-pricing system that: balances incentives for economic efficiency and investment; creates incentives for energy efficiency; balances development of the power system across the country; and promotes renewables. At the same time, China should improve methods for investment planning to better reflect real costs, and consider developing locational signals (via the auctioning of transmission capacity) as competitive markets develop.

Develop a cost-reflective approach to grid planning and investment. China should move away from its current "bottom up" and non-cost-reflective approach to grid planning. It should develop a transparent process for grid planning and investments that takes account of costs, as far as possible. In addition, the plan should ensure that energy-efficiency investments are properly considered as an alternative to supply-side investments.

Unwind subsidies; deploy new pricing, incentive and penalty mechanisms. To support cost-reflective pricing, China should start to unwind subsidies and cross-subsidies, and should increase the transparency of public funding for the power sector. These actions are steps toward eventually eliminating public funding. At the same time, it should continue to deploy time-of-use pricing and consider creating incentives and penalties that encourage consumers to improve their energy efficiency (inclining block prices, linking prices to efficiency standards for buildings via hook up fees, etc.).

Create lifeline support mechanism for poorer populations. In order to mitigate adverse social and distributional effects that often accompany tariff rebalancing, China should develop a lifeline support mechanism aimed at the poorer parts of the population. This lifeline must be carefully designed to be available only to those who really need it. Features of the lifeline should include effective targeting, positive net benefits, administrative simplicity, and transparency.

MANAGING DEMAND[39]

Making reform work on the demand, as well as the supply, side

Growing economies require growing infrastructures to meet increased energy demand. The power sector forms part of the general economy, as well as being a key input to the economy. Thus, it can be expected to continue to reflect the ups and downs of economic activity (Figure 8). Even if new infrastructure is well planned, growth of the power sector is not smooth, primarily because power comes in discrete units (newly commissioned plants, for example). Supply for a growing demand therefore shows the demand rising in a smooth curve, but the supply developing by steps. Smoothing the peaks and troughs of expansion requires a stronger policy focus on the demand, as well as the supply, side. This can only be achieved through

39. Specific recommendations related to this topic are given at the end of this section.

policies that will help to contain demand and allow demand response (IEA, 2003b). As demonstrated in Figure 8, annual growth rates in China's GDP, installed capacity and electricity generation, 1990-2005, reflect both the level of economic activity and the consequences of planning decisions.

Figure 8

Annual growth rates in China's GDP, installed capacity and electricity generation, 1990-2005

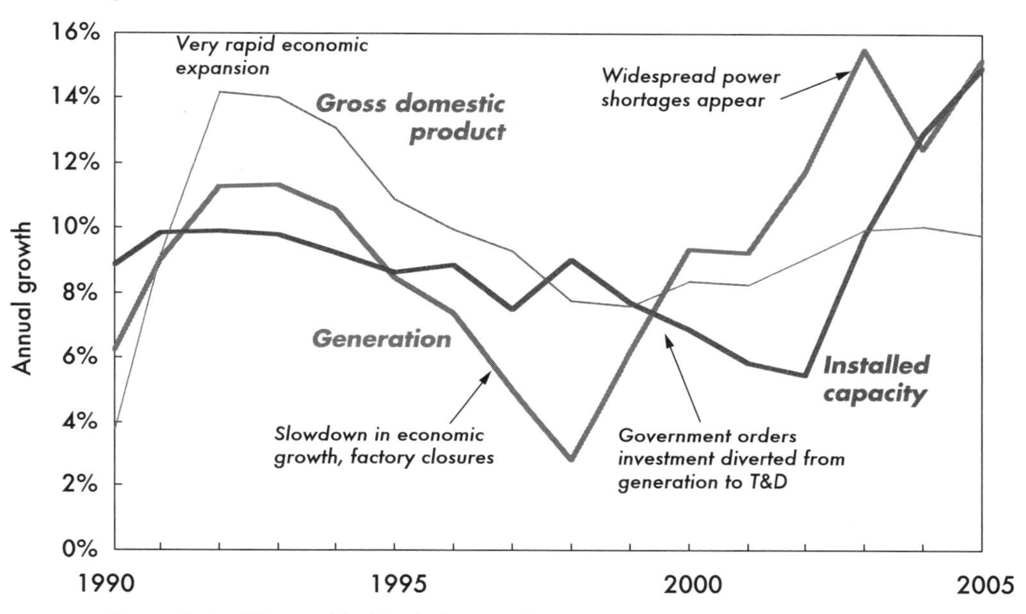

Source: National Bureau of Statistics (various years).

Investing in energy efficiency is often cleaner, cheaper, safer, faster and more reliable than investing in new supply. In addition to reducing the need to build new generation, transmission and distribution facilities, efforts to improve efficiency also reduce maintenance and equipment replacement costs, as many efficient industrial technologies have longer lifetimes than their less efficient counterparts. Efficiency measures (*e.g.* changing light bulbs, insulating buildings) can be implemented much more quickly than expanding energy supply.

From the perspective of the power system as a whole, the core issue that stands in the way of reducing energy use is how energy-efficiency improvements are valued against supply-side offers. A megawatt of power supply has a clear value to the power system and to power markets. From the perspective of balancing supply and demand, reducing power demand by a megawatt – or supplying a "negawatt" – results in the same effect, but it is typically not treated the same by the market. End-use decisions and supply-side decisions are made by various categories of players with different access to information and different objectives, even though they all directly impact the energy system. A fundamental goal of market restructuring is to put demand-side and supply-side decisions on an equivalent basis, and thereby to ensure that participants act in support of more rational use of resources. To help the reader better understand the concepts presented in this chapter, a series of definitions is presented in Box 12.

It is natural for electricity companies to build more generation, improve the grid, and trade electricity – supplying power is their core business. By contrast, energy efficiency in end use is a new area of business. In order to secure the development of energy efficiency, incentives – or even obligations – for energy efficiency must be directed toward companies. Such policies often come under the generic name of demand-side management (DSM).

Box 12

Definitions related to demand-side management

Energy efficiency and energy savings

Energy efficiency refers to reducing the total demand for energy for a given level of output – of goods, services, or economic benefit, *i.e.* deriving "more energy for less fuel". For the power sector, this means looking at energy efficiency, encompassing how energy is used in power generation, in transmission and grid management, and at end use. Its technical definition on the supply side is thermal efficiency, that is, the amount of fuel consumed to achieve a given energy output, or the rate at which the energy embodied in different fuels or electricity is converted into valuable products or services. In principle, improved efficiency should have positive impacts for economic growth, for the environment and for security of supply. Other things being equal, greater efficiency leads to a reduction in total energy demand, *i.e.* energy savings. However, often when efficiency leads to lower costs of producing or using energy, its consumption rises – the so-called "rebound effect". This is why a distinction must be made between energy efficiency and energy savings: the latter may need to be promoted in order to secure the benefits of the former.

A basic assumption behind policies to improve energy efficiency is that its benefits are a form of public good[40]. Improving efficiency provides widespread benefits that are not always or easily perceptible, so that companies or consumers that use electricity often feel no compelling interest in energy efficiency. Active intervention by government is therefore needed to promote energy efficiency and energy savings.

Negawatts

Energy efficiency in end use leads to energy consumption savings and, hence, reduces the amount of energy production capacity needed. If, for example, the amount of electricity saved is equivalent to the production of one new power plant, this allows supply to be maintained without building the extra plant. The use of

40. A "public good" is a commodity or service which if supplied to one person is available to others at no extra cost. This contrasts with a "private good" in which one person's consumption precludes the consumption of the same unit by another person.

efficiency savings to offset the need for new plant capacity is referred to as "negawatts". Negawatts are typically cheaper than traditional megawatts: they can be conceived in terms of "shadow" power plants that would have been needed if there had been no savings. Negawatts can be considered as one of the OECD's largest energy "suppliers".

Demand-side management (DSM)

Demand-side management (DSM) is a structured means of promoting energy efficiency and energy savings. DSM consists of a set of funded programmes aimed at promoting energy efficiency and energy savings in end use (by private consumers and industry, etc.). The goal is to reduce the market barriers that prevent companies and consumers from taking advantage of opportunities to improve energy efficiency, such as lack of information, funds, and incentives to save energy. DSM programmes can take a number of forms: *(i)* Financial incentives to end users, designed to modify energy use or change end-use equipment (*e.g.* switching to more efficient light bulbs or refrigerators); *(ii)* Educational programmes for end users on efficiency opportunities; *(iii)* Energy-efficiency performance contracts for companies; or *(iv)* Programmes to develop suppliers of end-use energy products and services, including energy service companies (ESCOs). DSM is mainly targeted at long-term energy savings (reducing load levels rather than load-shape management).

In many pre-market reform initiatives, DSM programmes were implemented by integrated utilities. However, DSM programmes have since faltered in many countries in the wake of power sector reforms. After market reforms take place, the DSM concept must be adjusted to take account of demand response from empowered consumers (a positive development) and fragmentation of market players (a major challenge for the implementation of DSM).

Integrated Resource Planning (IRP)

DSM is a part of a broader concept known as integrated resource planning (IRP). IRP targets the main strategic objectives for the power sector, *i.e.* environmental sustainability, reliability, and affordability objectives that embrace energy efficiency and energy savings. IRP promotes an integrated approach to the management of resources in the power system, aimed at meeting these objectives by assembling a mix of demand- and supply-side resources. For example, a healthy "portfolio" would consist of a diverse mix of power plants, as well as a mix of contracts and spot energy purchases, DSM investments and load management.

DSM and IRP: power sector management versus *market forces*

Some of the specific measures promoted under DSM may raise difficult trade-offs *vis-à-vis* policies to promote competitive markets and minimise social disruption from price rises. Energy-efficiency considerations imply the need to break the link between sales and revenues (power companies in competitive markets want to sell power, not save it). One suggestion to this end is capping generator revenue, which contradicts the principle of allowing prices to be set freely by the market. Adding a fixed sum to end-user tariffs (the "system benefit charge") in order to pay for energy-efficiency investments is also challenging, particularly in a context in which the aim is to minimise price rises.

More broadly, DSM and IRP imply a more managed approach to some issues, such as the mix of power plants, which would otherwise be left to the market in order not to interfere with market signals. To the extent that energy efficiency is a public good (and care for the environment an externality[41]), some adjustments are necessary. However, the right balance must be struck between the regulation needed to reflect these features of the energy sector and excessive management and regulation of power market players. It is also necessary to ensure that the relevant institutions (including power sector regulators) are working together with the bodies responsible for energy efficiency to resolve these issues.

Load-shape management: meeting short-term demand

Load-shape management involves reducing loads on a utility's system during periods of peak consumption, or allowing customers to reduce electricity use in response to price signals (demand response). Load-shape management seeks to ensure that the power system can cope with demand on a short-term basis, by shifting and smoothing short-term demand peaks, and raising elasticity of demand. Mechanisms for load-shape management include interruptible load tariffs, time-of-use tariffs, real-time pricing, and voluntary demand response programmes, as well as direct load control (IEA, 2005c). Demand response is an important new factor in the management of load shape in reformed power sectors. It can only happen in the context of competitive markets in which consumers are able to adjust their demand in response to changing prices for power (this is examined in more detail in the section on considerations for the longer term).

Load-shape management programmes are largely short-term responses and do not necessarily reduce demand on a long-term basis, though they can do so and hence help to reduce supply-side investment needs. They are not the same as load-level management, which consists of policies to reduce load altogether by reducing demand, *i.e.* saving energy over the long run. Both, however, are needed for an efficient and effective power sector, and they fulfil complementary roles.

The step-wise development of energy production in a fast-growing economy implies that the power system faces blackouts at times when energy demand is growing beyond the capacity of production to keep pace This normally triggers *ad hoc* action to cut off some consumers from power supply for a while, in order to reduce the peak to a manageable level. Blackouts are expensive because they are likely to lead to a loss of production and even a reduction of capacity in other sectors of the

41. A cost (or benefit) to society that does not have a market price attached to it, because it is not internalised (taken into account) in decisions taken by individuals and companies in the market. Society's welfare is diminished as a result.

economy. They are also a major source of disruption to comfort levels and welfare in the residential sector. Within the reform process, actions can be taken to mitigate or eliminate the occurrence such blackouts.

China already engages in a great deal of load-shape management. The concept has been deployed to handle system reliability in severe peak load situations, mainly in summer. For example the peak load reduction in 2003 was some 10 GW, of which only one-third was due to DSM. The remainder was triggered by explicit government orders, requests or recommendations in order to cope with emergency situations. The measures to obtain voluntary demand response included time-of-use pricing, which guided industry to rearrange production plans and make some load interruptible, and energy storage (shifting loads between peak and valley). Energy companies and their customers have also developed their relationships to facilitate load management.

China should consider whether it needs to implement a more systematic approach to load-shape management, by anticipating more clearly the peaks and valleys of demand. There are three main approaches to this:

- **Peak clipping.** In this approach, energy consumption is reduced at peak times, meaning that specific measures are taken that have an instant effect (Figure 9). In a well-managed system, this can be done with minimal impact on industrial production and residential comfort. Activities that can easily be stopped for a short time or at a particular time, *e.g.* water heating or air conditioning, may be targeted for peak clipping in advance of a crisis.

- **Valley filling.** This approach is pro-active in the sense of promoting adjustments in industry or personal behaviour to match times when demand for energy (*i.e.* when load or basic demand) is low. This could, for example, include night-time production.

- **Load shifting.** The final option discourages demand for energy at times when demand is known to peak, and encourages end users to shift to times when demand is known to be low. For example, industrial cooling could be discouraged at peak times, and washing clothes could be encouraged at times of low demand.

Figure 9 Typical load shape changes

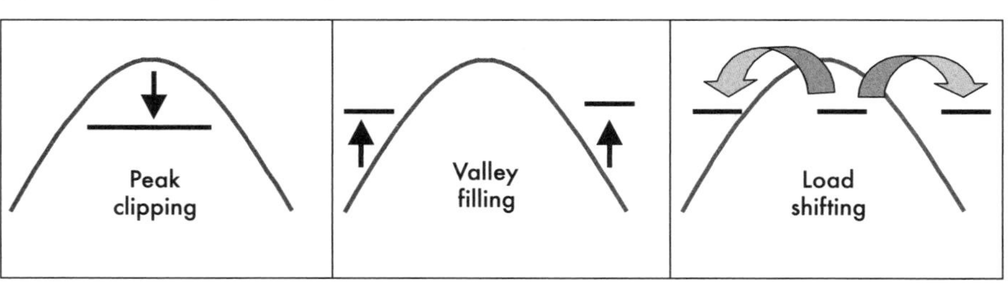

The benefits of a more systematic approach for load-shape management include: improved system reliability, by reducing peaks and improving safety margins; enhanced system security by reducing dependency on supply resources; and less costly network reinforcements, as energy-efficiency measures will be active alternatives.

Load-level management: reducing long-term demand

To date, China has placed too much reliance on load-shape management, to the detriment of policies aimed at reducing load. In addition, its regulatory framework is no longer adapted to supporting energy-efficiency improvements designed to reduce demand over the longer term.

There is still substantial scope to reduce energy intensity and save energy, both of which would hold a number of benefits for China. Resource constraints are one factor; although China is endowed with abundant coal resources, per capita energy resources are low. Reducing energy use would enhance security of supply and would also help to meet environmental objectives by reducing pollution. Improved competitiveness is a further potential benefit of more effective load-level management. The industrial sector is the dominant energy user in China, and energy makes up a major portion of industrial production costs. Ergo, reducing energy intensity through more energy efficient products makes good business sense.

China's energy-efficiency policies need new momentum

China has had policies in place to promote energy efficiency and energy savings since the early 1980s. Many are linked to its economic development plans and many have borne fruit. For example, conservation was incorporated into the national plan and major policies to support energy savings have included energy conservation planning, electricity saving regulations, financial and economic incentive policies (*e.g.* tax breaks for CHP; energy saving awards; subsidies for "green lights" and energy conservation projects); and the promotion of technologies for energy saving. The success of these measures over time is reflected in the significant progress in energy savings, which has, in turn, eased pressures on power production.

The 1997 Law on Energy Conservation is the framework under which China is developing a wide range of programmes, with the goal of reaching zero annual growth in energy consumption by 2040. The programmes include: promulgation of efficiency codes and standards; product certification and labelling programmes; development of ESCOs; energy-efficiency demonstration projects; training courses; and public education programmes.

However, there is still a long way to go in meeting best performance efficiency standards in power generation, transmission and distribution, and end use. To some degree, power sector reforms have halted the momentum: to date, the main focus has been on the supply side – primarily increasing generation capacity and, more recently, strengthening the grid. There has been no countervailing focus on the demand side and the promotion of energy efficiency. Worse, there has been a weakening of institutional capacities to advance energy-efficiency measures.

Before 1993, the energy conservation department of the then Ministry of Energy was responsible for managing power generation efficiency. It carried out tasks to improve generation efficiency, including reducing coal use per kWh, cutting individual power use, and reducing transmission and distribution losses. The affiliated National Electricity Conservation Office was responsible for end-use electricity conservation, which included the development of a plan to reduce consumption of electricity. The plan included several measures such as: limits for electricity use by various appliances; electricity consumption standards for equipment; monitoring of electricity use; and promotion of demonstration projects and advertisements for reducing electricity use. These institutions no longer exist, and their functions with regard to energy efficiency have been dispersed to other entities (some of which have also been replaced). There is no longer any single focal point for energy-efficiency promotion. Thus, what was once a strong body for adapting and integrating energy-efficiency policies for a competitive power market has become fragmented.

China's Tenth Five-Year Plan (2001-05) calls for the formulation of market-based incentive policies to promote energy efficiency, including tax policies and financial incentives, energy price reforms, and bank-lending policies. China's *Medium and Long Term Energy Conservation Plan* includes ten recommendations for the promotion of new market-based energy conservation mechanisms. These include, *inter alia*, power demand-side management, integrated resource plans, performance contracting, and several other DSM initiatives.

In 2002, the State Power Economic Research Centre and the Beijing Energy Conservation Centre submitted a report to the State Council entitled *Recommendations on Expediting the Promotion of DSM*. This was published as a decision reference by the State Council; it was the first time the Council stressed the role of DSM in China. Its recommendations included:

- Develop detailed DSM regulations to clarify the main DSM policies and roles of various stakeholders, especially government agencies and power grid companies.

- Establish a rational power tariff system, including time-of-use prices.

- Develop a DSM public welfare fund, based on the system-benefit charges adopted in other countries. The fund could equal 1-3% of customers' power bills and could initially be extracted from a portion of the urban surcharge currently added to power tariffs. The fund should be under the regulatory supervision of relevant government bodies.

- Make full use of existing organisations, such as the SETC Energy Conservation Information Dissemination Centre and the State Power Company's DSM Centre, to assist with recommendations, information and policy advocacy.

- Develop quality standards for high power-using equipment, adopt mandatory standards to phase out outdated equipment, and promote the extension of energy-efficient and technology-intensive products and equipment.

However, these proposals are not yet integrated systematically – or at all – in the plans for further supply-side reform of the power sector.

China needs a broad and sustained commitment to DSM programmes

China was first introduced to the concept of DSM in the early 1990s. Since then, government agencies, larger power consumers, research institutes, universities and other organisations have been active in its promotion. Activities to date include international exchange and co-operation, training courses, pilot studies, demonstration projects and educational activities.

Elements of DSM are now in place, but the picture is very uneven and there is no clear policy for its broad uptake. In 2003, SETC and SDPC issued a joint circular entitled *Announcement of Issuing Management Measures for Energy Conservation*. This recognised the importance of DSM as a power-saving strategy. It also included recommendations for local economic commissions to promote DSM, and for each provincial grid operator to study its situation for DSM and make proposals. As result, China has made significant progress in the development of ESCOs. Three such companies are located in Beijing, Lianoning and Shangdong. However, there are no regulations or policies requiring companies to invest in energy efficiency, or encouraging them to implement DSM to increase end-use energy efficiency.

Several DSM pilot studies have been carried out and show significant power savings potential and environmental benefits. However, none were ever implemented, except for a demonstration project involving peak load management (in Beijing), which is generally easier to apply than other DSM programmes. It seems from the work already carried out to test DSM that China is having difficulty moving from pilots to full implementation of promising ideas. What are the reasons for this? A review by the Energy Foundation (2003) of studies and tests carried out in Shenzhen (*Shenzhen Power Network DSM Pilot Study*, 1993), Shanghai (*Shanghai DSM Cost/Benefit Analysis*, 1994), Beijing (*Peak Load Management in Beijing*, 1996-98), and Jiangsu Province (*DSM in Jiangsu Province*, 2000-2002), suggests the following factors:

- Problems in defining the roles of different players (companies and layers of government) and difficulties in co-ordination.
- Lack of financing for the measures envisaged, linked to lack of incentives for companies to take part.
- Uncertain legal and regulatory environment.

The challenge of competitive markets: to sell or to save power?

Competitive markets have a profound effect on both the demand and supply sides of the power sector. Liberalisation leads to a fragmentation of the supply side, as the potentially competitive parts of the supply chain are unwound. In order to sustain competition, a core aim is to encourage a maximum number of players and prevent the re-emergence of a few dominant players. Players in a competitive market make money by selling power and seek to boost their profits, not reduce them. Energy efficiency is rarely core to, or even part of, the business of the new companies. In fact, owners may perceive that energy-efficiency improvements may reduce profits – or, at least, provide no obvious gain. Observing the dictates of the market, they would shed every expense possible – including demand-side programmes – to compete more aggressively. As yet, most new markets do not give clear or direct incentives for these individual market players to put resources into end-use efficiency improvements. Thus, under market liberalisation, demand-side energy efficiency may become even more of a public good, *i.e.* one that everyone benefits from but which no one stakeholder or group of stakeholders has a strong interest in providing. This is particularly true, if rules that previously required integrated monopoly utilities to set aside funds for demand-side efficiency no longer apply to the more numerous, newly competitive generators.

Congestion can be a powerful incentive for the players in a liberalised market, particularly in the sense of encouraging them to look for the cheapest way of avoiding it. However, the many players in a disaggregated market will not suffer congestion simultaneously, and this may lead to distorted decision-making. For example, a generator might wish to avoid running a temporary and expensive additional plant. But if this plant sets the price for the market, and the generator owns a second, cheaper plant, this creates an incentive to run the more expensive plant and harvest a "windfall profit".

The positive effect of competitive markets is that demand response can be used as a countervailing resource to supply options for managing load. Demand-reduction measures should be allowed to compete on equal terms with new generation sources. The two are, in fact, complementary to supply-side investments to strengthen the grid and interconnections. In a competitive market, conservation becomes a supply resource.

Securing a policy and regulatory framework to promote energy efficiency in competitive markets

It should be noted here that there is a wide spectrum of views on the place of DSM in fully competitive markets – and even on how DSM is best defined. DSM began as an approach that was adapted to pre-reform power sector frameworks. The need for, and appropriateness of, traditional DSM in fully liberalised markets should be reviewed once that stage of reform has been reached. In the meantime, care is needed to ensure that specific DSM programmes do not create obstacles to competitive markets. This could happen if the reforms focus DSM solely on those aspects of the power sector that one expects to be regulated indefinitely (the grid and system operation), so that the competitive parts of the market can develop their own approaches. It also generally implies a progressive move away from mandatory requirements and a stronger focus on incentives for voluntary energy-efficiency measures (including information campaigns). Demand response, which is only possible with advanced competition, can also be viewed as the next stage that replaces at least part of what DSM seeks to achieve, as consumers make their own supply/demand choices. All that said, one issue is very clear: China needs strong DSM policies now.

Developing DSM

China's reform creates a major opportunity to integrate DSM into the regulatory framework for competitive markets and into the regime for investment approvals. Ideally, this would aim at making DSM profitable for power companies; at minimum, it should remove the barriers and disincentives to DSM. Experience suggests that if DSM is to be part of a reformed power sector, it should be incorporated at the regulatory design stage. There are two reasons for this: *(i)* It will likely influence the structure of the reformed power sector; and *(ii)* It will be much harder to change the rules of the game once they have been established

(Box 13). The framework for DSM should include targets for specific activities, as well as overall targets for desired results in terms of energy-efficiency savings, which should be linked to a reporting and evaluation system. It is important to establish clear and stable institutional structures to take DSM measures forward.

Box 13

Lessons from other countries: the value of DSM in power sector reforms

Experience in a number of countries (including the United States, New Zealand, Chile and Argentina) shows that a competitive market does not automatically deliver energy efficiency and contain demand.

DSM programmes in the United States, as in many countries, faltered in the wake of power sector reform and restructuring that disaggregated previously integrated utilities. It also dealt a blow to the belief that market forces would be sufficient to promote energy efficiency. Investment in energy-efficiency programmes, not including load-management expenditure, declined dramatically from USD 1.6 billion in 1993 to USD 900 million in 1997. Much of this decline can be attributed to the elimination of regulatory requirements for utilities to conduct IRP and DSM programmes. Since then, however, government authorities have recognised the need to continue active measures and spending on DSM rose steadily to USD 1.10 billion in 2000.

California's experience

In 1996, before the power sector reforms went into effect, the California State government required utilities to invest in end-use power efficiency such as DSM. By 1999, California's energy-efficiency investments and standards had already removed about 10 000 MW from its peak demand, the equivalent of 20 large power plants. Overall energy-efficiency investment dropped by 40% when the requirements on utilities were removed, which contributed to a rapid growth in overall power demand and the resulting power crisis in 2001.

The subsequent reinstatement of energy efficiency and DSM programmes substantially reduced the economic and environmental damage associated with the crisis. California has increased funding for utility DSM programmes. It also extended, until 2012, the use of the "system benefit charge", a small charge on electricity bills to pay for continued investment in energy efficiency.

Source: Energy Foundation (2003).

Developing sources of finance for investing in energy efficiency

Competitive markets create a problem *vis-à-vis* energy-efficiency expenditure, *i.e.* figures that used to be buried inside integrated utility accounts are now exposed. In reality, both power companies and the grid in competitive markets still need capital to continue with energy-efficiency spending. Otherwise, it simply will not happen. There are several options for financing including:

- **End-user pricing**, which adds a small charge to consumer tariffs. The United States, Norway, Spain, Denmark and Thailand have adopted this approach. China already has experience with surcharges and this could, in principle, replace illegal surcharges. It is worth noting again that energy-efficiency improvements that flow through will ultimately reduce costs to consumers.

- **Grid tarification**, integrates the capital costs of energy-efficiency improvements in tariff calculations for the grid.

- **Capping generator revenue** provides a means of breaking the link between sales and revenue. It is, however, controversial.

The experience of other countries suggests the need to establish both incentives and mandatory requirements for investing in energy efficiency. Fortunately, there are a number of ways to do this. For example, performance based-regulation can be deployed. In this scenario, performance – in the form of meeting targets such as energy-efficiency spending – is linked to rewards for compliance. Some states in the United States negotiate annual efficiency targets with power companies and tie bonus payments to the achievement of these targets. For example, Texas requires its power companies to make energy-efficiency savings equivalent to 10% of each year's growth in power demand. Some Australian states (where energy intensity is high and energy efficiency low due to cheap electricity) have licences for the supply and distribution of power that require companies to develop and implement DSM and environmental strategies. "Informative billing", which includes information on the development of energy consumption and options to reduce that consumption, should be considered to encourage all consumers to save energy.

Promoting energy-efficiency "aggregators" to counter fragmentation

To counteract the fragmentation that currently exists within the power sector, China should seek to re-build stronger relations amongst market players. ESCOs can play a key role in this area. They are valuable in delivering efficiency to institutional and large commercial markets and especially useful in helping large customers improve their energy use efficiency. ESCOs can emerge successfully out of power companies, via affiliates. They do have limitations as their deployment might leave 85% of the market uncovered (residential, small commercial and industrial customers), and they examine only the customer's bottom line, not the systemic impact of energy use. DSM funding programmes have played a major role in creating and supporting ESCOs.

Distribution companies are another main potential focus for aggregation activities. They can be required, for example, to investigate whether demand-side alternatives are more cost effective than building new grids. To implement this requirement, distribution companies would need to obtain customer-specific retail sales data and monitor/verify DSM results. Australia (via a codified regulatory principle that alternatives to investment be considered in the context of a cost-benefit assessment) and some US states have adopted this approach. Norway allows distribution companies to recover only a part of their investment in new network capacity, which provides a financial incentive for them to examine alternatives to expansion.

Specific DSM activities

DSM activities and advice on energy savings should seek to cover all major fields of consumption in which important savings potential can be identified, thereby building on what China is already doing in areas such as:

- Promoting energy-efficient electric motors, lighting and household appliances.
- Providing advice to energy intensive industries, tailored to each industry.
- Undertaking activities to reduce energy consumption in existing buildings, including thermal insulation and the use of energy efficient building components.
- Undertaking activities to promote efficient heating systems and to reduce losses in existing installations.
- Undertaking activities to reduce the need for cooling and air conditioning and to ensure the use of efficient systems.
- Providing advice to service companies linked to office buildings.

Box 14 The IEA's Implementing Agreements (IAs) and Chinese participation

The IEA's Implementing Agreements provide the basis for interested parties to collectively undertake energy technology research, development and deployment activities. IAs are supported by a system of standard rules and regulations that allow interested member and non-member governments to pool resources and conduct research to support the development and deployment of particular technologies. There are now more than 40 such collaborative projects with several thousand participants from 58 countries, organisations or companies working in the areas of fossil fuels, renewable energies and hydrogen, energy end-use (transport, buildings, industry, etc.), fusion power, and cross-sectional activities.

China is currently party to two Ias – Small Hydro, Fusion Materials – and is a sponsor of the IEA Clean Coal Centre. All three of these deal with technologies that in the short or long term would strengthen China's electric power sector:

- The Chinese party to the Small Hydro IA, which is intended to provide objective, balanced information about the advantages and disadvantages of hydropower, is the Hangzhou Regional Centre (Asia-Pacific) for Small Hydropower.
- The Ministry of Science and Technology is the Chinese party of the Fusion Materials IA, which is aimed at developing data for the international tokamak (magnetic confinement fusion) experiment (ITER), planned to be built in France, and other fusion design activities.

- The Beijing Research Institute of Coal Chemistry is a sponsor of the IEA Clean Coal Centre in London, which provides state-of-the-art information on efficient and environmentally sustainable use of coal worldwide.

Chinese representatives have also expressed interest in joining several other IAs, such as the Hybrid and Electric Vehicles IA. More information on IAs can be found at http://www.iea.org/ia/list.aspx.

China's potential participation in the IEA's Implementing Agreement on DSM

The IEA oversees a mechanism, known as an Implementing Agreement (IA) under which like-minded parties in both IEA member and non-member countries can collaborate in a variety of energy technology areas (Box 14). Activities of the DSM Implementing Agreement span a wide range, including:

- DSM programmes in the changing electricity business environment.
- Communication technologies for demand-side management *e.g.*: "Flexible Gateway" technology for information to flow in and out of homes.
- Innovative procurement of demand-side technologies.
- Techniques for implementing demand-side management in the market place.
- Demand-side bidding in a competitive electricity market.
- The role of municipalities in liberalised systems.
- Performance contracting.
- Energy use, metering and pricing for demand management delivery.
- Demand response resources.

Recommendations for managing demand

Strengthen load-shape management. China should seek to strengthen its approach to load-shape management through a more systematic deployment of options that have the capacity to smooth out the peaks and valleys of demand.

Create legal framework for demand-side management. China should secure an appropriate legal framework for the comprehensive development of demand-side management (DSM) activities. The framework should include provisions for:

- Financing DSM activities.
- Promoting investment in energy efficiency through incentive-based regulation, as well as mandatory requirements.
- Establishing measures to promote energy-efficiency aggregators.

- Conducting a review of the institutional structures for promoting DSM and establishing these structures on a clear and stable basis.

- Integrating DSM and demand response into the regulatory framework for competitive markets and investment planning.

Participate in international DSM activities. It would be highly beneficial for China to join the IEA's Demand Side Management Implementing Agreement (DSM IA). Current members of the DSM IA actively encourage China's participation, both to share IEA country experiences with China, and to develop a procedure for sharing experiences between China and other countries.

TOWARDS EFFECTIVE MARKETS: ACTIONS FOR THE NEAR TERM[42]

This section sets out the steps that China should take now to strengthen the framework for the roll out of competitive markets, and includes proposals for a modest start to developing some limited competition across the country. The main arguments and issues are summarised in Box 15.

Box 15 Near-term steps towards fully competitive markets

One of China's most pressing needs is to secure a reliable power supply. This creates a compelling case for starting to undertake some carefully planned near-term action towards competitive power trading across the country. Other elements of this report underline the importance of actions to manage demand as well as to strengthen infrastructure (especially of the grid). But there is scope to squeeze more out of current infrastructure without necessarily having to build more – the key is to make more efficient use of the existing infrastructure. Realising this "efficiency dividend" involves the development of incentives, which are currently lacking, for more efficient behaviour. Specifically, gains could be achieved through system dispatch that applies the economic merit order and through generation companies that are encouraged to pay more attention to their efficiency (and perhaps less to growing their market share). This implies a move away from China's current single buyer approach.

A significant gap exists between current and potential utilisation of installed generation capacity. Capacity utilisation of current power plants may be around 55-60%; for many of them it could be closer to 90%. (A significant portion, of course, such as hydropower stations limited by seasonal water flows, cannot approach such high capacity factors.) Very broadly, this implies that more efficient capacity utilisation may have the potential to deliver the equivalent of 200-300 GW of new capacity. In Australia, more efficient capacity utilisation is estimated to have accounted for roughly two-thirds of the total efficiency gain related to market reform in the early years.

42. Specific recommendations related to this topic are given at the end of this section.

Unlocking the potential for more efficient capital utilisation

In order to gain more power from existing capacity – and begin setting the stage for competitive markets, China should undertake several key steps in the near term. The top priority is to strengthen the framework for competitive trading, which requires two specific actions. The first step involves securing full grid and system operation independence. This is the only way to secure dispatch that is soundly based. The second step involves securing independent and efficient generation companies. These companies need to be put in a position where it is in their best interests to look for efficiency gains in order to survive, *i.e.* where they can no longer rely on state funds if they encounter difficulties. Together, these steps would have the immediate effect of improving the efficiency with which existing capital is used.

The next step, which could be developed simultaneously, is the careful introduction of some limited competitive trading between different regions and provinces. Competitive transactions could start between the grid companies and the largest consumers, who would be allowed a choice of supplier. To accelerate the move away from single-buyer model, the grid companies' purchasing function should be clearly separated from their transmission function. This could evolve into transactions between generation companies and large consumers.

Other key elements of strengthening the framework include:

- Pricing that is more cost-reflective across the whole value chain, and that includes grid pricing as a separate element.
- Stronger institutional capacities to manage and regulate the emerging competition.
- More efficient corporate governance, of both grid companies and generating companies.
- Transparency in terms of information and data, at all levels, so that the regulator, market players and consumers can know how the market is operating.

Advancing these key steps requires careful fleshing out. For example, it will be necessary to address the handling of existing contracts for power supply and to establish non-discriminatory access to the grid. However, these are feasible near- to medium-term goals.

Reforming state domination of China's power sector

Despite the diversification of ownership and widespread partial privatisation through stock market flotations, China's generating sector remains dominated by the state, albeit at different levels of government and through different types of enterprise[43]. The government's approach to ownership of the power sector (and the wider energy

43. Early reforms in the mid-1980s opened up generation to investment by third parties outside central government. These were mainly provincial and local governments, but they also included private sector interests, both foreign and domestic. Power plants built and purchased by these investors now account for more than half of total capacity. They are sometimes called independent power producers (IPPs), but this is misleading: most remain intimately linked to the government, and are not really independent.

sector) is not the same as for many other sectors of the economy. Under the economic reforms of the last two decades, a wide range of industries has been privatised and loosened from state control. However, the power sector remains a strategic sector, not least to ensure control over energy supplies. The government is also keen to promote the growth of large corporations that can compete on the international stage.

To date therefore, there has been no full privatisation of a major state-owned energy company in China. In almost all cases, the capital opening of energy companies at different levels of government has resulted in the state retaining a majority of the shares, which are often non-tradable. These ownership stakes may be held either directly by the state, or by domestic legal entities that are themselves owned by the state. There are no plans for full privatisation of the five main generators or the two grid companies, and it seems unlikely that this policy will change significantly in the foreseeable future. The government remains the main supplier of capital to the power sector.

Independent grid and system operation ensure fair system dispatch

Grid independence is essential to reassure the market that system dispatch will be fair. If an incumbent utility retains control of the grid, or the grid company retains an interest in generation, it can easily limit or even exclude access to the grid by competing generators. Without non-discriminatory, third-party access to the grid, new generators will not invest in new capacity. China has laid a sound groundwork for further reforms to strengthen the grid and system operation. Grid ownership and management have not been fragmented, and China has opted to separate grid ownership from generation. By creating just two grid companies, the potential for these companies to discriminate against power supply from other grids is less than if a larger number of separate companies had been created.

System operation is a vital function in competitive markets, yet it should also be fully independent of the market itself. The two crucial roles of the system operator are to secure reliability and fair competition. Electricity cannot be stored (economically), and the grid needs to be kept in balance at all times. Therefore, the system operator must balance supply and demand across the whole grid, by managing the interface between the market and actual physical outcomes. It is a natural monopoly, delivering a form of public good, and needs to make the best decisions in relation to plant dispatch for the market as a whole.

That said, system operation needs to be regulated in order to secure independence, and this is a key role for SERC. Once competition starts to develop, the system operator is not only the market manager but also a potential player in the market. A market-based system contains "grey zones" in which the system operator may be expected to be involved. For example, managing congestion may require the purchase and sale of electricity to eliminate bottlenecks. It may also be necessary to procure and manage ancillary services, which may also require buying and selling power at short notice. The need for the system operator to take action drives the need for regulation. Moreover, it is impossible to prescribe rules for every interaction that might be encountered by this function. Therefore, system operators need discretionary powers.

At this stage of the reform process, the independence of China's grid and system operation is not yet assured. The grid companies, which also control plant dispatch, retain an interest in generation, and generators as a whole retain close financial links with the state at different levels. The structural reforms that have allowed the development of generation companies need to be completed, by fully detaching generation interests from the centrally owned grid companies. This needs early attention; even the perception that generators are still affiliated to the grid – and therefore might receive favourable treatment – will affect competition and the willingness to undertake new investments. SG's "non-core" business (thermal and hydro construction companies, etc.) should be divested. Inadequate unbundling from grid interests compromises both balanced system development and commercial efficiency. It may, specifically, compromise transmission investment in favour of generation investment, if inadequately unbundled grid owners are left to make their own choices. It also increases the difficulty of market entry.

The 2001 OECD Council recommendation on structural separation in regulated industries addresses the need to separate potentially competitive activities from regulated utility networks. It also considers the need to guarantee access to essential network facilities to all market entrants on a transparent and non-discriminatory basis. The recommendation notes that, in the absence of anti-trust or regulatory controls, incumbents have both the ability and the incentive to restrict competition, and that this generally harms efficiency and consumers. Incumbents can, in particular, cross-subsidise competitive from non-competitive activities. Commercially sensitive information can also be made available between different company activities, which can advantage the incumbent's competitive activities relative to those of other players.

Improving governance of grid and system operation

This report assumes that the grid will remain state-owned for the foreseeable future. This means that separating government's role as owner is especially important for China. SG is an SOE owned by central government. Although the establishment of SASAC is a major step forward in shifting the government away from direct interference in the management of SOEs, the government continues to be responsible for the nomination, assessment and dismissal of not only the CEO, but also of senior executives. State interference in management is likely to continue. At the very least, procedures for recruitment of top management in the grid companies should be transparent and based on objective criteria of relevant business competencies.

For the longer term, there is a need to establish a clear corporate structure for the grid and system dispatch (this issue is considered in more detail in the next chapter). Two different approaches to securing an effective independent governance framework for the grid and system operation have emerged from reform experiences elsewhere: the ISO (Independent System Operator) model (separate system operation from grid ownership) or TSO (Transmission System Operator). The TSO is simpler as regards governance and co-ordination, but raises issues of independence. The ISO is more likely to be independent but raises issues of co-ordination. The TSO helps the co-ordinated development of regional markets. However, a main argument against combining system operation with grid ownership is the difficulty of regulating such a powerful company. The regulator needs strong powers to deal with this.

The TSO route would appear to be the simplest for China, and is in effect being developed now. China, like many other countries, started out with a single vertically integrated entity. Developing a TSO out of this is simpler than making a more fundamental change by separating system operation from grid management. A TSO is also better able to raise capital, implement projects and make timely decisions on grid expansion. On the other hand, the ISO approach offers some important advantages. The size and power of the grid companies, combined with the weakness of current regulatory oversight, provides an argument for splitting the two functions to minimise the risk of market dominance by a single entity.

The importance of independent, well-governed generation companies

Effective and modern systems of corporate governance are critical to the efficiency of a company. The structure and organisation of a company, its management systems, decision-making processes and operational practices all have an impact on commercial efficiency. In the case of China, stronger corporate governance would be better able to tackle the issue of corruption, primarily by reducing the opportunities for rent-seeking and dissipation of revenues. Robust corporate governance is even more important in the context of state ownership, where countervailing forces for inefficiency are at work, and market mechanisms to discipline enterprises or contest ownership and control are weak – or cannot function at all.

The measures needed to strengthen corporate governance, especially where state ownership still prevails, are often known as "corporatisation". This means, broadly, that all activities are organised in well-defined companies with clear ownership, responsibilities and objectives. A number of countries have deployed corporatisation, including various Nordic countries and several Australian states, in order to strengthen the efficiency of power companies and to ensure that they act independently in competitive power markets.

Corporatisation helps to address three key issues: transparency, efficiency and competition. Transparency in the use of public funds is important for the efficient development of China's power sector. At present, this transparency is virtually non-existent. There is, for instance, considerable uncertainty regarding the actual performance of generating companies. The rise in coal prices that generators have had to absorb has put many under considerable pressure, as has been widely reported, but generally without precise details[44]. At the same time, some generators continue to grow market share and may be generating super-normal profits. Generators in difficulty can easily negotiate *ad hoc* support from local governments, often in ways that are not clear beyond the circle of parties to a particular deal. Either way, the situation is a liability for the healthy development of competitive power markets, as well as the sound development of local economies. Corporatisation also tends to align the interests of managers with greater economic efficiency. So long as there are concomitant moves towards greater cost-reflectiveness, this would also be a step towards promoting greater energy efficiency. By setting normal procedures for market entry and clarifying ownership rules, corporatisation is also an important means for encouraging the generation sector to develop and diversify through new entrants and ownership.

44. One estimate is that in the first two months of 2005, the profits of power companies were one third lower than in the same period of 2004. More generally, the *OECD Economic Survey – China* (2005) reports that over 35% of SOEs are not earning a positive rate of return, and one in six has negative equity.

Despite considerable progress made in recent years, China still suffers from weak corporate governance. This progress includes the creation of SASAC, the introduction of new accounting practices, and the restructuring and enhanced transparency that has followed from capital opening. Comprehensive changes to company law are being drafted, which will place all businesses, regardless of their type of ownership, on a comparable legal and regulatory footing, at least in principle. China is moving toward international best practices in this regard.

But much remains to be done. Generally, corporate governance remains weak in China, both in the private and public sectors. Outstanding issues include:

- Many state-owned assets have not yet been incorporated, and therefore lack boards and other essential internal governance structures.

- Where boards exist, they tend to focus on the enforcement of government regulations rather than on the corporation's long-term goals.

- Since only a minority portion of listed SOE shares can be traded, participation by institutional investors is relatively low. In addition, disclosure by listed companies is limited, sanctions for inadequate disclosure are weak, and limited protections for minority shareholders leave scope for abuse by controlling shareholders (typically the state).

- The implementation and enforcement of current laws is a large issue.

- The state controls key enforcement bodies, such as the stock exchanges, and the judiciary is weak.

- Although China does have incentives that encourage top management to succeed, these are distorted by the absence of a strong corporate governance framework. Typically, salary, bonus and share option schemes allow senior management to share in the profits of a successful enterprise. However, the current weakness of corporate governance can result in senior management taking an undue share of the profits or otherwise dissipating them to the detriment of the company and the shareholders.

- Incentives to avoid failure are not well defined. In a market economy, the penalties for failure of a privately-owned company include dismissal of the senior management, bankruptcy or take over. None of these mechanisms are applied in a transparent and consistent manner to China's SOEs.

The five large generating companies' internal practices still derive from the old State Power Company, of which they were an integral part until the end of 2002. There is an absence of relevant internal systems, appropriate performance measurement systems, authorisation systems for risk management and investment, and transparent internal control and audit systems. Management and staff culture also need to develop.

Detaching generating interests from local economic development is a priority. At present, generation companies are often too closely linked to local governments, and this raises problems. Provinces are reluctant to increase power imports if this means that local sources of generation will be displaced. Companies are offered soft budgetary terms, and know that they can count on rescue in case of difficulty. Companies operating under such conditions have no incentives to build competitive strength because of lack of financial pressure. Protection leaves company management with very little real understanding of the real risks facing the company, and no experience of how to manage these risks. It is also damaging to competitors because it makes the playing field uneven.

A number of steps can be taken to secure generators' genuine independence and better corporate governance. These include:

- Effective and transparent unbundling of generation accounts from the accounts of other state interests to which they are attached (Box 16).
- Clearer and more robust corporate governance rules and controls.
- The implementation of rules to secure a neutral framework for competition.

Sustainable progress also requires broader reforms of the relationship between the centre and local governments, particularly as a means to minimise inclinations to use the power sector as a support for local economies.

Box 16 Lessons from other countries: clear accounting and transparency needed in use of public funds

In the absence of full divestiture, clear accounting information is needed to secure effective unbundling and promote transparency in the use of public funds. This reinforces the importance of "fit-for-purpose" regulatory accounts as a means of enforcing effective separation. Regulatory accounts differ significantly from ordinary financial accounts. In the first place, regulatory accounting principles were developed to establish a clear separation between competitive and monopoly aspects of the value chain in previously integrated utilities. The same principles are just as relevant for separating utilities from their state owners.

The following principles were developed by the Independent Regulators Group (IRG), a consortia of European national telecommunications regulatory authorities:

- **Regulatory accounting principles.** These principles should establish the key doctrines to be applied in the preparation of regulatory accounting information. They should include, *inter alia*, the principles of cost causality, objectivity, transparency and consistency.
- **Methods for attributing costs, revenues, assets and liabilities.** A description should be given of the attribution methodologies used to fully allocate revenues, costs, assets and liabilities.

- **Basis for transfer charging.** A description of the basis used to transfer charges between different parts of the entity should be given, as required under the accounting separation rules. Typically, this will prescribe methodologies for ensuring that an entity charges itself on the same basis as other entities for similar services.

- **Accounting policies.** These should follow the form used for the preparation of standard statutory accounts and should include, for example, details of fixed asset depreciation periods. Where the regulatory accounts are prepared on a current cost basis, the basis on which the assets are valued should be included.

- **Long-run incremental cost (LRIC) methodologies.** If LRIC applies, a description of the methodologies used to prepare long-run incremental cost information should be given. It should include details of the identification and treatment of shared or common costs.

The IRG emphasises that "financial information prepared and published for regulatory purposes often differs significantly from other financial information prepared by companies for statutory or other purposes" and that "the basis on which regulatory accounts are prepared requires special regulatory rules as well as the application of generally accepted accounting practices". They also note the value of procuring an independent audit opinion on the accounts, "which enhances the quality, objectivity and credibility of the information presented".

Source: Independent Regulators Group, http://irgis.anacom.pt/site/en/

Developing clearer and more robust corporate governance rules and controls is important to all electricity reform initiatives and China is no exception. The OECD (2005a) notes that the transition to a more efficient system of control would imply a number of measures (Box 17)[45]:

- Creating and enhancing the role of boards in SOEs.
- Improving the recruitment and performance evaluation procedures for senior management.
- Strictly separating the government's exercise of its ownership in SOEs from its regulatory and other functions.
- Eliminating interference in SOE management.

45. This OECD report also notes that a revised company law is under consideration. This should aim to improve corporate governance, notably offering better protection to minority shareholders in both quoted and unquoted public companies, and defining the role of corporate bodies such as the supervisory board and the duties of directors.

Box 17 The 2004 OECD Principles for effective corporate governance

The OECD *Principles of Corporate Governance* were originally endorsed by OECD Ministers in 1999 and updated in 2004. The Principles include guidelines which are grouped under six headings: *(i)* Ensuring the basis for an effective corporate governance framework; *(ii)* The rights of shareholders and key ownership functions; *(iii)* The equitable treatment of shareholders; *(iv)* The role of stakeholders; *(v)* Disclosure and transparency; and *(vi)* Responsibilities of the Board.

The preamble includes the following:

"Corporate governance involves a set of relationships between a company's management, its board, its shareholders and other stakeholders. Good corporate governance should provide proper incentives for the board and management to pursue objectives that are in the interests of the company and its shareholders and should facilitate effective monitoring. The presence of an effective corporate governance system helps to provide a degree of confidence that is necessary for the proper functioning of a market economy. As a result, the cost of capital is lower and firms are encouraged to use resources more efficiently, thereby underpinning growth."

Of particular relevance to China is the principle on disclosure and transparency:

"The corporate governance framework should ensure that timely and accurate disclosure is made on all material matters regarding the corporation, including the financial situation, performance, ownership, and governance of the company." These include issues such as company objectives, financial results, governance structures and an independent annual audit. A strong disclosure regime that promotes real transparency is a pivotal feature of market based monitoring of companies. Disclosure also helps improve public understanding of the structure and activities of enterprises, corporate policies and performance with respect to environmental and ethical standards, and companies' relationships with the communities in which they operate.

Source: OECD (2004).

Ensuring a level playing field for private interests

Boosting private sector engagement also implies the need to ensure that private investors are not discouraged by an uneven playing field. Issues that need attention include:

- **Initial balance sheet.** The structure of the balance sheet used by publicly owned entities at the time competition is introduced ultimately affects their basic cost structure and, hence, their competitive position in the market. If assets shown on the books are substantially undervalued, and if debt and equity positions do not conform to private sector norms, the State entity enters the competitive market with a built-in competitive advantage over private sector rivals.

- **Pension and other liabilities.** The above scenario is true for pensions and other liabilities.

- **Taxation.** If their tax regime is not reformed prior to competition, publicly owned entities may enjoy unfair tax advantages over competitors. Typically, public entities often start with some tax exemptions.

- **Separation from the public budget.** The budget needs to be transparent and protected, so that public authorities do not have to choose between funding the electricity public service and wider government interests.

- **Internal subsidies and cross-subsidies.** At the very least, effective accounting separation must be put in place to prevent cross-subsidisation of competitive activities (generation for example) out of revenues from non-competitive activities.

- **Rate of return on assets.** Public entities need to recover their costs fully, including an appropriate rate of return on capital (neither too high nor too low).

- **Public guarantees.** Public guarantees – *i.e.* mechanisms that make the public authorities liable if the publicly owned entity cannot meet its debts – lower the risk attached to their borrowing, compared with a private company. Public guarantees, explicit or implicit, should be unwound as far as possible.

Towards competitive power trading

Leaving aside the pilot-organised power markets in Northeast, and East, and Southern China, most of China's power market remains under a system of government-directed power sale transactions, with the grid companies as single buyer. Generation is sold under long-term contracts, set and approved by the government, to the grid companies. In turn, the grid companies sell the power to end users, again under government approved retail tariffs.

Elsewhere in the world, trade between jurisdictions has often proved to be an effective lever for developing competition and improving efficiency. China could take modest steps now to move away from the government-controlled single buyer approach and develop some competitive trading:

- In a first step, the grid companies could be allowed to establish their own (initially modest) power sale transactions (*i.e.* not directed by government) with large consumers. This would involve giving the largest consumers freedom of choice (a freedom which could be extended, incrementally and over time, to smaller consumers). This step is best taken once the grid companies are detached from generation interests and have achieved a greater degree of independence, though it could be started now.

- In a second step, direct transactions between generators and large customers could be allowed, *i.e.* no longer using the grid companies as single buyers.

- To be effective, this would require the grid companies (and hence system dispatch) to be independent of generating interests. Trading mechanisms for the allocation of transmission capacity to underpin the transactions could be developed (*e.g.* basic auctions). Such mechanisms would not only help develop competitive transactions, but they would also help to identify bottlenecks in the grid.

Strengthening the market infrastructure for competitive trading

Spreading private ownership, including foreign ownership

Across the Chinese economy as a whole, private sector dynamism has offset the initial negative impact of down-sizing public enterprises, leading to new sources of employment. The sharper incentives directed toward private sector companies have led to higher productivity from a lower use of capital and labour. As a result, the aggregate productivity of private companies in the industrial sector is estimated to be almost twice that of enterprises controlled directly by the state. State-owned companies' behaviour is not like that of private investors. They are more interested in scale than efficiency.

A useful way of encouraging positive change toward greater efficiency in the generation sector is to spread ownership as far as possible. Spreading ownership by technology, location and size, as well as by type of company, would help to inject new ideas and boost reform – from the reform of corporate governance to the adoption of new technology. Encouraging ownership that is diversified outside the power sector may be especially interesting, as such companies may not respond to price signals in the same way as those for whom power is the core business. Giving more headroom to the private sector is likely to raise efficiency and release government capital for other uses. Increasing the stake of the private sector and giving state institutions with holdings in generation the right to transfer their holdings to the private sector would also help.

There is a particular need for more foreign investment in China's power sector. The relative lack of foreign investment over the last ten years, alongside a massive rate of domestically sourced investment, raises the question of whether China's power sector requires foreign investment and, if so, why (Box 18). In principle, a number of reasons may be identified:

- To increase the amount of capital invested.
- To shed some of the investment risk.
- To gain access to foreign technology.
- To enhance the technical skills of the workforce.
- To gain access to foreign management skills.
- To introduce competition to the incumbents.

The recent behaviour of banks and power companies suggests that the first two of these rationales are not currently applicable in China. This may change if pressure increases on the banks to apply more rigorous criteria in their evaluation of loans, or if a surplus of power supply emerges, which discourages domestic investors.

Foreign technology is available and China's power companies have the funds to buy the best technology in the world, should they wish to do so. Further, many of the world's leading manufacturers are in China already – making the components for power stations and, in some cases, constructing the plants. This also provides for the transfer of the technical skills in the workforce.

The last two rationales for foreign investment are most relevant today. China's power sector requires advanced management skills for two reasons. First, it is in the national interest for all parts of the power sector to be managed with a higher level of technical efficiency. This will improve the supply of power to end users and constrain the demand for primary energy, as well as reduce the level of a wide range of pollutants. Second, it will be in the interests of the newly corporatised power companies to improve their financial and technical performance ahead of the planned introduction of competition. These objectives can be achieved through the range of joint venture and build-operate-transfer arrangements which were introduced during the 1990s, as well as through wholly owned foreign ventures. A further benefit of direct foreign involvement is the competitive pressure that would arise and would oblige all players to continuously seek to enhance their technical and financial performance.

Box 18 Foreign investment in China's power sector

Foreign investment in China's power sector has failed to reach the levels anticipated by the central government and by investors themselves. In the mid-1990s, the government estimated that an investment of some USD 10 billion per year was needed in order to raise generating capacity from 200 GW to 300 GW by the year 2000, and that 20% or USD 25 billion of this would come from overseas. The actual level of foreign investment was less than half of this, and almost completely stopped in the late 1990s, as central government imposed a moratorium on the construction of new large-scale power stations.

The total installed generating capacity surpassed 500 GW in 2005, and will reach 600 GW in 2007. This massive rate of investment since 2002 has been achieved with very little direct foreign investment. The continuing policy uncertainty concerning the nature of the planned power markets, combined with the unwillingness of the government to approve power purchase agreements, have succeeded in keeping all but the bravest of foreign investors away. At the same time, China's state-owned banks have been prepared to lend the required sums to the power companies. It would appear that both the power companies and the banks are prepared to accept the policy risk, on the basis that either the growing demand will secure the revenue flows, or that they will be protected if the loans cannot be repaid (OECD, 2003).

The need for a regulatory framework for foreign investment

The Chinese government has consistently provided tight guidelines for the involvement of foreign companies in the electric power sector through the

Catalogues for the Guidance of Industries for Foreign Investment. The Catalogue, issued in 2002, continues the approach of earlier Catalogues, explicitly forbidding foreign involvement in transmission and distribution, and limiting involvement in coal-fired power stations to those with a single-unit capacity of 300 MW or greater. Foreign investment in power stations using clean-coal technology, natural gas or renewable energy is encouraged. Nuclear and hydroelectric power stations remain open to foreign investment, subject to a Chinese controlling interest.

Despite this official "encouragement", substantial foreign investment in China's power sector is unlikely to be forthcoming until the nature of the rules for the new power markets are clarified and until effective regulatory and institutional arrangements are made to support these markets.

Ensuring that generation licensing does not obstruct market entry

An effective licensing regime is essential to ensure a fair and efficient market for new, as well as existing, players. Licences may be divided into two types: licences for the construction of new plants, and licences that set out the rights and responsibilities of the plant owners. China's licensing arrangements remain complex and the licensing process is often slow[46]. This provides opportunities for corruption and can result in increasing costs for companies. An over-complex licensing environment slows decision making and puts companies off from investing, which is especially damaging in a context of rapidly growing demand for power. A "one-stop shop" approach for licensing would help.

Policing anti-competitive behaviour

Although two important documents – *Methods on Power Market Supervision* and *Basic Rules for Market Operation* – were issued by SERC in 2003, related implementation is still at a preliminary stage. A number of regulatory issues need SERC's attention, in order to ensure that markets develop effectively.

At this stage, the most important issue is the need to establish a framework, implemented by SERC, to police and deal with anti-competitive behaviour, which has already started to emerge. A major issue for the development of power markets in China is the ability to identify and act against anti-competitive behaviour. All power markets show a marked tendency in this direction (Box 19), but there are specific factors in China that are likely to make the task of managing anti-competitive behaviour even more challenging. The first is local government support for local power companies, and a general climate in which generators are more inclined to put their efforts into better bidding strategies than into cutting costs. The second is regional market dominance by one or more of the large five generating companies. The third is that China has no experience of regulating anti-competitive behaviour and lacks a competition authority. The fourth is that a weak grid is likely to lead to congestion as trade develops, and generators may try to take advantage of this to get higher prices. Finally, demand is basically higher than supply, which further encourages generators to abuse their market power.

46. The World Bank (2000) noted that licensing may require approval from as many as 11 agencies at several different levels of government.

Evidence of difficulties already exists. According to some observers, the Northeast China market is not yet functioning properly because bid prices rarely move[47]. The market share of the five "new" state-owned generators was designed to provide a competition-friendly balance. But it has not lasted, as companies have sought opportunities for growth and exploitation of market power. Initially, each individual company was limited to holding a 20% share of total capacity within its region. However, there has been no clear regulation limiting these companies to this market share as new generation capacity is added. Indications suggest that share of generating capacity by some generation companies is growing quickly and already exceeds the 20% limit. In addition, nearly all the new generation capacity approved by NDRC since 2003 is owned by the five unbundled companies.

Box 19 Lessons from other countries: sustaining competitive power markets

Sustaining competitive power markets requires an in-depth understanding of their inherent market power. Market power can be defined, in general terms, as the ability of a seller to reduce the output supplied to the market so as to raise the market price, and to do so profitably. Examples of the abuse of market power are agreements to raise prices, or to create artificial shortages.

Electricity markets are prone to market power (Hunt and Shuttleworth, 1996; Hunt, 2002). They have special features that make them particularly vulnerable. Because of its physical characteristics, electricity cannot be stored economically in large quantities. It is subject to a wide variety of demand conditions, and yet the amount of power supplied to the grid must equal the amount taken out at all times in order to maintain electrical equilibrium and avoid physical damage to the grid. Short-term demand is highly price inelastic, and supply becomes highly price inelastic once physical capacity constraints are approached. Each point in time represents a distinct product market, and for each product market only some generation sources will be able to meet marginal demand. Due to limited storage potential, transactions cannot be easily substituted through time. The potential share of residual demand (demand that remains to be met after all plants but one are running at full capacity) controlled by each generator is very important. A generator's ability to control residual capacity within each region can put it in a monopoly position.

Legally, abuse of market power is a matter of determining that a company with a dominant position is abusing its position to realise substantial and sustained profit. But proof of abuse is difficult to obtain in power markets because it is extremely difficult to define the relevant market and product. Standard market concentration measures do not capture the situation.

47. The earlier experiments with power markets were also challenging. In the Zhejiang power market, it was observed that "market prices reach the price caps in nearly 10% of hours, bidders routinely meet and discuss their experiences in the market, and most, if not all bidders, follow the same bidding strategy of withholding about 10% of their plant capacity, which is bid at the cap price". (The Regulatory Assistance Project, 2003). Regular meetings and co-ordinated bidding would violate competition laws in most countries.

Tools and techniques of competition analysis

The usual competition tools that serve other sectors need to be adapted to take account of electricity's special features. However, they can be used as a baseline for developing approaches to counter the problems described above. Competition policy analysis suggest six key areas for attention if market power is to be constrained: *(i)* Increase the scope of the product market; *(ii)* Increase the scope of the geographic market; *(iii)* Increase the sensitivity of demand to price; *(iv)* Decrease concentration among existing suppliers within the relevant markets; *(v)* Increase the size and sophistication of customers; and *(vi)* Reduce barriers to entry.

Evidence of problems in other markets

The lesson of other reform experiences is that generators will exploit any scope for exerting market power that they can find, often through abuse of dominance or merger activity. The level of competition remains a serious concern in many liberalised markets, which are tending back to high levels of market concentration. Post-reform rationalisation of European power markets is a striking example. Between 1998 and 2002, the EU market saw 96 major mergers and acquisitions. By 2002, seven large utilities dominated this market, controlling nearly two-thirds of sales.

There are three main approaches to tackling such market power problems:

- **Engaging the competition authorities** (if they exist). Competition in the Nordic markets is monitored and regulated by competition authorities in all Nordic countries. This task can be shared with the regulator.

- **Ongoing monitoring.** In the United States, the Pennsylvania-New Jersey-Maryland (PJM) Market Monitoring Unit[48] produces a comprehensive annual State of the Market report, which includes various Herfindahl-Hirschman Indices (HHIs) that serve as concentration measures, as well as analyses of the special circumstances in which one plant is a pivotal supplier.

- ***Ad hoc* enquiries.** The EU Commission has launched a sector enquiry on electricity market competition in the EU power markets. The focus is the functioning of wholesale markets and price formation, including levels of market integration and the functioning of cross-border trade, as well as relations between grid operators to examine barriers to market entry.

48. The PJM Market Monitoring Unit is responsible for safe and reliable operation of the unified transmission system and for the management of a competitive wholesale market across the control areas of its members. It oversees the functioning of the market, which includes: assessing the state of competition in each of the PJM markets, identifying specific issues and making recommendations, monitoring compliance with market rules, and issuing an annual report on the state of the market.

Avoiding national fragmentation in the development of regional markets

Over the last two years, China has taken the important and ambitious step of setting up pilot regional power markets. This echoes the way in which power markets have developed elsewhere. For example, in Australia competitive power markets started in the state of Victoria (1994), and were then extended to New South Wales (1996), followed by Queensland, South Australia and the ACT (1998)[49]. It is a good way to test new skills, such as risk management, and system operation under conditions of increased trade, both for companies and regulators. In reality, there are a number of ways in which markets can evolve, as shown by other regions and jurisdictions around the world (Box 20) (various market models are described in more detail in the following chapter on *Considerations for the longer term*). It is not too soon, however, to think about how the evolving regulatory framework and market rules may influence the longer term development of China's power markets, *i.e.* whether they will lead towards an integrated market, or a series of essentially independent, coupled markets.

Keeping options open: regional power markets may set stage for unified national market

China has taken the only approach possible at this stage in its decision to develop regional power markets. It is too large a country to establish a single national market from the start of the reform process. The markets that it has started are at least equivalent in size to the Nordic European market, or the PJM market in the Northeastern United States – in other words, large. More broadly (leaving aside the pilot markets), the Chinese power sector is to some extent already modular, consisting of a two-tier structure of regional markets and grid companies responsible for system operation and real-time balancing, and a national level for inter-regional electricity trading. In addition, the dispatch control of generation resources is at the regional, not national, level. This structure is rather different from the single, centralised market models such as the PJM model. A centralised PJM-type market model would require the dissolution of regional dispatch centres and merging these into a large national grid/system operator. That said, a PJM-style centralised dispatch model might be adapted on the regional level and designed to co-exist with a national decentralised market.

There are, on the other hand, a number of arguments for sustaining a strategy that will allow the eventual emergence of an integrated national market in China. In terms of strengthening competition, the general concept is: the larger a power market, the better. China does not have any realistic prospect of linking up with competitive power markets in other countries in the foreseeable future. Thus, it has to foster its own competition. China also suffers from regional imbalances in the distribution of its energy resources and in economic development. These factors underpin its current strategy of developing the grid from west to east. Optimising resources across different regions suggests not only the further development of a unified grid, but that power markets should also tend toward unification. Provincial/regional markets should be aligned as soon as possible with efforts to create inter-regional links, which would pave the way for a national market. This evolution can be promoted by regulation.

China has an advantage over other jurisdictions that have developed competitive power markets. Being closer to a unitary than a federal state structure (at least in

49. Tasmania will be fully incorporated into the National Electricity Market in the near future.

principle) gives the central government stronger levers to develop integrated country-wide policies. This contrasts, for example, with the federal structures of the United States and Australia. However, reforms so far have tended (unwittingly) to reinforce provincial demarcation and raise inter-provincial barriers to trade.

Box 20 Lessons from other countries: dealing with market fragmentation

The EU experience – admittedly in the context of separate countries – gives a clear flavour of the difficult issues that could emerge if power markets and system operation go their separate regional way in China. It has taken many years of effort to develop a single EU market for electricity and there is still a long way to go. The 1996 directive (EU law) started the process; it took another seven years for a second, stronger directive to be approved. A fully integrated market is still work in progress. The key lesson from this is that market fragmentation is hard to reverse.

The EU experience of encountering difficulty in efforts to promote co-operation across the separate EU system operators[50] may be an argument for developing a single system operator in China. The EU Commission is seeking, with great difficulty, to develop more integrated and competition-friendly ways of dealing with the allocation of grid capacity and cross-border congestion management. A recent law seeks to improve transparency regarding information on capacity, as well as to better define how the available capacity should be allocated. It is suspected that system operators are currently managing congestion within their respective systems so as to push it to their national borders. Complex ideas such as flow-based market coupling, which seeks to bring together the separate markets, are also under development to address the issue of market fragmentation.

In the United States, reforms and the development of a unified market have been handicapped by the split between federal and state regulatory jurisdiction, as well as by the state action doctrine under which federal authority extends only to cross-state wholesale trade. To this day, the United States do not have a unified market, only regions that have come together. Australia – another federal country with significant power at state/territory level and with a similar split of responsibilities – has also had to battle and negotiate for the development of its National Electricity Market (NEM).

System operation is the glue, but common rules are needed

System operation is the core of an approach that will secure eventual unification of power markets. This can either be a move toward unified system operation, or the development of close co-operation between separate system operators. Australia has set up a single national operator (NEMMCO). The United States' PJM market

50. The fact that system operation is combined with grid ownership (the TSO or Transco model) is also relevant, as the (national) grid owners are tempted to manage system operation in their own interests, and not in the interests of the single EU market.

has also developed according to the principle that market extension implies the expansion of unified system operation. By contrast, the Nordic market continues to have separate system operators, but an association (Nordel), which seems to work well, has been established to reach agreement on rules for trade and cooperation. The European Transmission System Operators (ETSO) functions in the same way in the EU, but with limited success.

Some variation between regions will continue to be needed in order to accommodate differences in the pace of reform and readiness for change (*i.e.* there is no need for all parts of China to have retail competition at the same time). But some elements need to be common from the start, and the national regulator should be empowered to approve variations in order to ensure that there are no impediments to future market integration. Common features can be steadily developed, and the experiences of the first markets can be used to define common rules that will help development both within each market, and between them. Key common features are likely to include a uniform bidding platform and common basis for transactions, as well as consistent wholesale and grid-pricing concepts across regional/provincial boundaries.

Setting a framework for system security[51]

System security, energy security (fuel inputs), and adequacy of investment (in the grid and generation) are the three pillars of security of supply (IEA, 2002). System security means the ability of a power system to withstand the unexpected loss of key components. It is centred on the transmission grid and system operation. Transmission provides the link between those who generate electricity and those who consume it. System operation ensures the short-term balance, security and reliability of the system.

Although it may seem far off at this stage, China's power sector reforms can be expected to have the same effect as elsewhere, which is to increase power trade and change patterns of grid usage. Anticipating these developments is important, in order to avoid unpleasant surprises in the future, but also to support system security now (Box 21).

There are both strengths and weaknesses in the configuration of China's current power system. It is helpful that structural reforms to date have established only two grid companies. This allows for integrated management and oversight of system operation, a major potential strength compared to more fragmented jurisdictions. The decision to unbundle the grid from generation interests is also a major step towards securing independent system operation, a key component of system security. However, the grid infrastructure remains weak. Even allowing for further planned investment, it will face challenges in coping with increased trade when competition takes off, which makes an effective framework for system security even more important. System operation in China would also benefit from investment in improved technology, as well as human resources.

51. This section is based on IEA (2005b).

A key aim of power sector reform is to stimulate competitive forces by engaging a multitude of market players in increased trade. These developments also create more dynamic system operating conditions. A new framework for system security is needed to deal with the new conditions.

Box 21

Lessons from other countries: competitive power markets require stronger system security

Power market reforms typically engender more efficient use of transmission systems, and greater regional integration of power flows resulting from inter-regional trade. In turn, greater integration helps to improve overall transmission system security by permitting more effective reserve sharing. But growing demand for transmission capacity to accommodate inter-regional trade means that transmission systems are increasingly run at or near their security limits. The unbundling and decentralised decision making that comes with power market reforms also implies that many decisions that were once centrally co-ordinated within vertically integrated utilities are now made by many independent market players. As a result, previously stable and relatively predictable patterns of network use have, in many cases, been replaced with less predictable usage, more volatile flows and greater use of long-distance transportation, reflecting growing inter-regional trade.

The new patterns of transmission network use create a more complex and dynamic operating environment. Real-time monitoring and management by system operators becomes increasingly crucial for maintaining system security. At the same time, unbundling reduces system operators' capacity to manage system security through co-ordinated actions across the value chain. The emergence of regional markets that span multiple control areas has added to the challenge by increasing each system operator's exposure to the operational decisions of other system operators – and to potential failures beyond their area of control.

Blackouts linked to system security frameworks that failed to keep up with market reforms

Market liberalisation and the disaggregation of vertically integrated utilities in the EU have required a fundamental reappraisal of the institutional pillars on which European regional electric system security rests. Prior to market reform, each utility managed its own control area, both in terms of generation and transmission. This meant that a single party was usually able to ensure compliance with security requirements. The absence of competition, coupled with generous implicit or explicit provisions for returns on investment, also helped to promote a high level of voluntary compliance with security rules. The 2003 blackout in Italy highlighted a number of defects in the European framework for assuring system security[52]. The blackout was not caused by some extraordinary event such as a storm or terrorist attack, but by weaknesses in the legal, management and technical frameworks for assuring system security including:

52. The underlying factor was increased trade between France and Italy. Loop flows in Europe's highly meshed grid took some of this trade through Switzerland, where the crisis started.

- Weak co-ordination and information exchange between system operators.
- Lack of legally enforceable responsibilities, and regulatory frameworks that were based largely on industry self-regulation.
- Inadequate tree trimming along power lines.
- Application of the N-1 system reliability standard[53] in a deterministic way, which did not take account of the probability of a failure occurring or the impact of potential failures.
- Slow and inadequate system operator responses.

The 2003 North American blackout also highlighted the inadequacy of existing voluntary reliability rules, especially lack of enforcement. The Energy Policy Act 2005 has since given the United States' federal regulator (Federal Energy Regulatory Commission) authority to appoint a body to implement and enforce a set of mandatory security rules.

An integrated policy response needed to create a new framework for system security

The unbundling of the old utilities within market reform raises a key question: which agency should be responsible for system security under the new conditions? Responsibilities need to be clearly allocated and co-ordinated among the many new stakeholders: system operators, transmission owners, regulators, and market participants. In order for this to happen, governments need to provide strong policy leadership on a number of related issues that need attention:

- **Legal and regulatory framework.** Obligations imposed on the old utilities are overtaken with market reform and must be replaced. Responsibility and accountability for system security need to be re-allocated among a number of players, and secured under a new and enforceable legal framework. National grid codes that set out harmonised technical and operational requirements are an effective way of enshrining the rights and obligations of grid users and system operators.
- **Security standards.** The way in which the N-1 security standard is interpreted and applied should be reviewed to take account of the probability of a failure occurring and the impact of potential failures.
- **Co-ordination, communication and information exchange.** Operating practices need to reflect the more dynamic environment, and allow real-time responses to system emergencies.
- **Investing in technology and people.** System operation can be considerably enhanced with the use of better technologies, which can improve the accuracy, quality and timeliness of information. It can also support the development of more

53. The N-1 system reliability rule, used worldwide (with regional variations), states that the system must be operated in such a way that any single incident, for example the loss of a line, should not jeopardise the security of an interconnected system.

dynamic system modelling and more effective control of power flows. Investing in people is equally important. Highly trained and experienced personnel are needed to manage system security under the new conditions.

- **Asset performance and maintenance.** The reliable performance of transmission assets is fundamental to system security. This starts with the establishment and enforcement of improved maintenance standards and practices.

- **Vegetation management.** Contact with trees is one of the most common causes of transmission line failure. Again, this means a review of current standards and vegetation management plans.

Recommendations for moving towards effective markets: actions for the near term

Separate generation interests from grid companies. Plans to complete the full detachment of generation interests from the grid companies should be completed, as soon as possible.

Improve corporate governance. China should review the corporate governance framework for the State Grid Corporation and the China Southern Power Grid Company Limited, with the aim of minimising state interference in the management of each enterprise. In addition, China should consider implementing OECD (2005a) recommendations on corporate governance, tailoring them to the power sector. This includes activities such as:

- Creating and enhancing the role of boards in state-owned enterprises (SOEs).

- Improving recruitment and performance evaluation procedures for senior management.

- Strictly separating the government's exercise of its ownership in SOEs from its regulatory and other functions.

- Eliminating interference in SOE management.

Implement "regulatory" accounting. Under SERC's management, China should establish a framework by which generation companies could produce and monitor transparent "regulatory" accounts.

Unbundle generation and state accounts. China should take steps to effectively and transparently unbundle generation accounts from the accounts of other state interests to which they are currently attached. In addition, rules should be developed to secure a neutral framework for competition in the generation sector, particularly between private and publicly owned players.

Expand competition beyond pilot markets. China should consider taking some modest, first steps towards the development of competitive power trading outside the pilot markets that are already established. A first step might be to allow grid companies to establish their own transactions (not directed by government) with large consumers (this might best be carried forward once grid companies are fully separated from generation interests). Subsequently, direct transactions between generators and large consumers could be allowed.

Encourage private investment – domestic and foreign. China should strengthen the policy and regulatory framework to encourage independent domestic and foreign investment in power generation. As competitive markets develop, it should ensure that these independent power producers have the third-party grid access necessary for carrying out transactions.

Develop mechanisms to manage anti-competitive behaviour. In the absence of a competition authority, China should pay special attention to strengthening the regulatory framework for managing anti-competitive behaviour. It should take action for the rapid development, implementation and enforcement of market rules that promote transparency and a comprehensive flow of information on market operations. The rules should be included in an instrument with legal status, under SERC's supervision.

Build flexibility into system operation and market rules. China should bear in mind the future possibility of a unified, country-wide power market. Thus, in the development of system operation and market rules, it should avoid developing multiple regional frameworks that would be difficult to integrate at a later stage. Effort should be made to identify those elements that need to be common from the start: a uniform bidding platform and common basis for transactions; consistent wholesale and grid pricing concepts across regional/provincial boundaries; etc. At the same time, SERC should be empowered to approve proposed variations in order to ensure that there are no impediments to future market integration.

Strengthen system security: China should act now to strengthen its framework for system security – rather than waiting until increased trade makes this a more urgent issue. Elements that require attention include:

- The legal and regulatory framework.
- Security standards.
- Co-ordination, communication and information exchange.
- Investing in technology and people.
- Asset performance and maintenance.
- Vegetation management.

TOWARDS EFFECTIVE MARKETS: CONSIDERATIONS FOR THE LONGER TERM

As China progresses through power sector reforms, it is worth considering the characteristics that distinguish an effective competitive power market. The most fundamental element is that a competitive power market is one in which electricity is traded among generating companies, intermediaries (such as traders and distribution companies) and end customers (preferably both large companies and smaller customers). Power is traded either bilaterally between the market players or through bidding in organised markets, or both. A number of contractual arrangements underpin the trade, including short-term bidding through the spot market, futures and forward financial transactions, and trade in longer-term contracts. This implies a major shift away from the single buyer model, currently in place in China, under which generators cannot sell power directly to customers, but must sell instead through the original monopoly incumbent or grid company that manages the transaction.

In keeping with its long-term target of a more competitive market, China can derive confidence from the evidence that shows such markets do work – provided that a strong regulatory framework is in place (Box 22). In particular, an effective competitive market must give market players the opportunity to make decisions according to the signals of an undistorted market price. The lesson emerging from other reformed markets is that price signals – if left on their own – will do their job. Several cases provide evidence that government interventions to cap prices below what can be justified by economic reasons can blur price signals and slow market responses. Contrary to some perceptions, the blackouts that have affected certain jurisdictions can be largely blamed on inadequate and outdated system security frameworks. In other cases, potential blackouts have been avoided because the market was able to respond to tight supply by reducing demand. But the regulatory framework needs to be robust and well designed for this to work, and there must be no political interference once the government has "released" the market.

Box 22 Lessons from other countries: price signals work in competitive markets

Price volatility and high peak prices are inherent features of well-functioning competitive power markets. Effectively managed, they provide essential information to those market players that can – and need to – respond. Price caps to contain volatility are unhelpful; they blur the price signal on which the market needs to act. Evidence shows that there are better ways to manage the effects of volatility. For power companies looking to hedge risk, financial future markets offer one solution. For consumers wishing to insulate themselves from unpredictable power bills, fixed-price contracts can be the right choice.

Demand response can also help to mitigate price volatility. The ability of consumers to see and react to spot market prices is key to encouraging demand response. If, for example, prices in the market vary hour by hour, but customers do not receive such signals (if they are billed, for example, only on a monthly basis) they do not have the opportunity to reduce their consumption when hourly prices spike at high levels. If they could respond in this way, it would actually dampen price volatility. Even though many customers prefer to stay with a fixed price, only a proportion needs to practice demand response in order to affect the market. For response to be an option, however, they must have access to real-time prices.

The Nordic experience

The liberalised Nordic market experienced a supply shock in 2002. A severe drought in the autumn of that year depleted hydro reservoirs in the region. As a result, production corresponding to 15% of Norway's consumption and 9% of overall Nordic consumption "disappeared". The situation was exacerbated by a cold autumn that pushed up power demand. The market response was four-fold: electricity spot prices quadrupled; thermal power production in the region increased (mothballed plants were brought back on stream); imports to the Nordic market increased; and, not least, consumption decreased. Consumption in Norway fell by some 5%, the main reductions being from large industrial consumers (who were also spurred by an economic recession). But households and companies with electrical boilers for industrial process heating also contributed.

Strong pressures for government intervention – including a major public debate on the issue – were resisted; instead, governments reiterated their continued support for the market. The longer term effect was also noteworthy. Prices stayed high but average prices declined and there were no more spikes. Sweden also reduced consumption in response to the crisis, although households did not contribute to this as they typically had 1-2 year contracts to insulate them from short-term price changes.

The Australian experience

Australia, which also has liberalised power markets, experiences "needle peaks" most summers in the states of Victoria and South Australia. This occurred again during the first three years of operation under the newly formed NEM. Again, the government decided not to intervene. Investors responded by making significant investments in peak capacity: 450 MW of gas turbines in South Australia, and 550 MW of gas turbines in Victoria.

The California and Ontario experiences

In contrast, government intervention in North American markets had disastrous effects. In California, price caps on wholesale markets prevented generators from passing through cost increases to consumers and were a direct cause of the 2001 crisis. The Ontario government capped prices at an early stage of market liberalisation. This sent a signal of no confidence to the market, and led to pressures on government to reverse the reforms.

Source: IEA (2005b).

Factors for successful competitive power trading

A number of factors need to be present for successful competitive power trading. In its transition to competitive power markets, China will need to consider how it will create or develop:

- A robust regulatory framework, including carefully designed market rules, to clarify rights and responsibilities, and to ensure that market participants act fairly and transparently. This needs to be embedded within a broader framework of an effective legal system and judiciary.
- Effective grid access and fair system dispatch.
- Transparency.
- Liquidity, which implies a large enough number of players and transactions.
- Spot trade as a significant proportion of overall trade, as the spot price acts as a guide to efficient longer-term contracting.
- Significant demand-side participation from a variety of consumers and intermediaries.
- The existence of financial markets, especially contracts for differences (CfD), which allow generators to hedge risk and hence mitigate price volatility.
- Effective ongoing market policing, to identify and take action against anti-competitive behaviour.
- Adequate physical infrastructure, notably an excess of generating capacity over demand, and effective transmission and distribution infrastructure.

The following sections consider five broader issues in more detail, specifically: grid and system dispatch, third-party access to the grid independence, a competitive market structure and demand participation, transparent and robust market rules, and choosing among various power market models.

Reinforcing grid and system dispatch independence

China, like many other countries, started out with a single vertically integrated entity which owned the physical infrastructure (power plants, grid) and carried out all the functions necessary to deliver power to the end user (generation, system operation to ensure the reliable operation of the grid including deciding which plants should run and when, and planning for and investment in new infrastructure). A competitive market requires greater separation between these functions. At this stage in the reform process, China has a broad choice in the model it will select to increase independence. The underlying question is: Should grid ownership be combined with system operation?

Two different approaches have emerged from reform experiences elsewhere. The independent system operator model separates system operation from grid ownership. The model of a transmission system operator model keeps these two functions together (Box 23). Both have strengths and weaknesses that must be taken into account. As discussed in the previous section, *Actions for the near term*, given its history of a single, vertically integrated power market, the TSO route would appear to be the simplest for China, and is, in effect, being developed now. On the other hand, the ISO approach has its own merits, which should be considered.

Box 23

Lessons from other countries: effective governance and regulation key to system operation

In general, two models exist for system operation: the transmission system operator (TSO or Transco) and the independent system operator. In reality, both have strengths and weaknesses, and there is no clear winner between the two models. The choice of model is often dictated by initial market structures and ownership, and by political feasibility. For example, the United States and Argentina developed ISOs because grid ownership was spread across multiple (private) owners. By contrast, in much of Europe, TSOs have emerged from the original vertically integrated and publicly owned utilities. Regardless of the approach, there is a need to confer the system operator with input to long-term grid planning.

The Transmission System Operator

A TSO may be under private or public ownership. It is a regulated "for profit" corporation that owns and operates all transmission facilities within a given geographic area. An example is the United Kingdom National Grid Company.

The strongest argument for a TSO is that it keeps grid asset management and planning under a single roof. In principle, it is best that all grid-related activities – planning, operation, investment and maintenance – are conducted within an integrated framework. This facilitates effective management of the trade-offs between short-term system operation and network access, and long-term investment and planning.

The downside is that transparency suffers. It is a challenge to create governance structures that promote transparency and establish incentives that lead to unbiased operational and investment decisions. For example, the TSO may favour increasing grid capacity to meet growing load at a particular location, even if new generation is a cheaper alternative. As TSOs are usually formed out of the original monopoly utility – *i.e.* they are typically what remain when generation has been spun off – some of the old monopoly instincts continue. This is especially true if the TSO also remains state owned, which may slow the process of market opening. It is likely that the slow process of developing the EU single market in electricity is due, in part, to adoption of the TSO model in restructuring the sector. TSOs in the EU market have problems with independence and have only muted incentives to develop an integrated market.

The Independent System Operator

The ISO is usually a non-profit entity that operates but does not own the grid. It has leasing or transmission control agreements with the company or companies that own the grid. Tariffs to cover the capital/operating costs of the grid are collected by the ISO and remitted to the grid owners. The ISO may charge a grid management fee to cover its operating costs.

An example is the Australian National Electricity Market Management Company (NEMMCO)[54]. NEMMCO is a company, separate from the grid-owning companies, formed under company law with shareholders comprising the governments of each participating jurisdiction in the Australian federal system. It is responsible for operating the national electricity market in accordance with the National Electricity Code (a legal instrument).

Transparency and neutrality are the strong points of this approach. It is easier to ensure that system operation is truly independent and an ISO is more likely to make unbiased decisions about grid investment and expansion. Separate grid ownership also makes it easier to develop a competitive framework for grid investment (keeping in mind that grid investment is an alternative to generation investment). However, there is a co-ordination issue; it is a challenge to create governance and regulatory structures that secure effective co-operation between system operators and grid owners. These relationships have proved to be complex. Congestion management is especially difficult. If the system operator does not own the transmission assets, the regulatory framework must provide the asset managing system operator with incentives to invest adequately.

Whichever model is adopted, and whatever the ownership, three key elements are critical to successful system operations:

- Effective structural unbundling from generation. This is especially important if the grid and generation remain state owned or maintain close interests to the state.
- Firm regulatory oversight. Overseeing the TSO/ISO is a key responsibility of the regulator.
- Sound governance structures. Independence requires careful attention to appointment procedures, etc.

Broadly speaking, a well-regulated public grid company – which has been firmly separated from generation and other competitive interests and is kept at arm's length from its government owner – will be more effective in supporting the emergence of a well-functioning power market than a privately-owned grid that has not been effectively unbundled from generation.

54. These institutions have now changed to the Australian Energy Market Commission (AEMC) and the Australian Energy Regulator (AER), which have replaced NECA and the ACCC. The ACCC remains responsible for competition supervision.

Model selection warrants serious consideration

The question of whether power market management should be combined with system operation should not be taken lightly, nor should a decision be allowed to "happen" by default. It must be taken consciously in the context of the broader decision of whether to confirm SG and CSG as TSOs (combined grid owners and system operators), or to detach system operation from grid ownership. In principle, separating system operation from market management is not a good idea because important synergies are lost (Box 24). If the TSO route is confirmed, this would add to the potential power of the two grid companies, and would require correspondingly robust regulatory supervision to ensure that they act fairly in the market. Therefore, the question for China is: Can such effective regulatory supervision be secured?

In reality, China's eventual decision is about whether the functions of market management and system operation should be carried out by one entity. This decision needs to be made in the context of the broader decision whether to go for a TSO or ISO model and the strength of the regulatory framework to manage the entities that emerge.

Box 24

Lessons from other countries: combining system operation and market management

Although they have very different mandates, there is good reason to favour combining the functions of system operation and market management. Three arguments are particularly relevant:

- Coupling the two functions covers the pricing of ancillary services such as reserve power. As the system operator must be involved in managing imbalances, this has the merit of simplicity (at least relatively speaking) – which is likely to enhance efficiency. In contrast, keeping the market separate from system operation is likely to require disaggregated procurement of balancing services, which increases complexity.

- It simplifies the task of congestion management.

- Integration with the market may promote a more efficient network management, making it more responsive to the needs of competitive electricity markets.

There is some evidence to suggest that keeping the two functions separate may optimise prospects for the power market to be run commercially. The system operator may be able to focus more on the technical demands for maintaining system stability, rather than on promoting an effective market.

Most countries have opted for combining the two functions (including United Kingdom, Pennsylvania-New Jersey-Maryland, Australia and the Nordic market). Spain is one exception, but co-operation is an issue between the system operator and the market operator.

Regulating grid access and use

Grid users need the security of evidence that the grid and system operation are managed independently. They also look for a clear set of rules covering both access to, and use of, the grid and want to be assured that a regulatory body exists to monitor and enforce these rules. China does not, as yet, have any rules for access to, and use of, the grid. It will need them once competitive power markets start to evolve beyond the current single buyer model. The responsibility for this regulation should lie with SERC and its lower level equivalents (*e.g.* regional/provincial regulators).

For China, the arguments in favour of regulated access to the grid are even stronger than usual (Box 25) – and more likely to result in fair outcomes. Despite institutional weaknesses – *e.g.* regulation at the local government level, links with local power companies – regulated access will be preferable to negotiated access, not least because of the great negotiating power of SG. If construction of grid capacity continues apace, it might be possible to limit access to unused grid capacity rather than having to seek a "fair share" of existing capacity. The construction of transmission capacity in China has consistently lagged behind the growth of generating capacity, and transmission constraints continue to be a major contribution to power shortages. If this continues, "fair" access may still face local institutional and political obstacles. Given the current problems associated with regulatory weaknesses, the lack of a competition authority, and a weak judiciary, China might want to consider the Nordic approach, which combines elements of regulated and negotiated access.

Box 25

Lessons from other countries: regulated access is preferable to negotiated access

The way in which electricity users access the grid has implications for competitive markets. As the term implies, negotiated access means that users must barter with the grid company for access to, and use of, the grid to support their transactions. Regulated access infers that there are rules for access to and use of the grid, which are automatically applied, in a consistent fashion, when a user seeks access. Regulated access is now the norm in virtually all OECD countries, but it is valuable to consider the differences.

The downside of negotiated access is that it runs the risk of considerable delays, particularly if a dispute arises. Negotiated access usually means (though not always) that there is no regulator to take up issues and disputes are referred to the courts, often via the competition authority. In reality, the courts are usually overburdened and often do not have the expertise to assess the situation. Thus, the process of reaching a settlement is lengthy, and users may give up before a case is concluded. Market entry suffers in consequence. Germany recently moved from a negotiated to a regulated approach, establishing, in the process, a regulator that replaces the role of the previous competition authority. The move followed complaints that the *ex post* management of access via the competition authority created significant delays and impeded market entry.

Regulated access removes any aspect of negotiation from the transaction process; the grid company is obliged to implement the rules. The regulatory approach does not avoid litigation altogether; regulators can expect to be challenged. However, once the courts have settled an important test case, it is likely that other cases will be simpler to handle. Clear rules expedite dispute resolution; without them, every case risks becoming a complicated affair. Effective enforcement of the access regime is critical, which relates to the main disadvantage of regulated access: it is more complex and resource intensive for the regulator. One way of minimising potential problems is to follow the example of some Nordic countries, which have set up an *ex post* access regime that relies on the application of an *ex ante* methodology.

Developing a competitive market structure and demand participation

Separating distribution from the transmission grid

The effective unbundling of networks is crucial to the creation of a competitive market framework. Separating distribution from the grid enhances the prospects for effective competition in power markets, as well as the prospects of stronger investment in this part of the infrastructure.

These are strong arguments for separation in the not too distant future. Conceptually, the distribution function is very different from the function of transmission grid management. Distribution companies perform a variety of functions such as meter reading and billing, and maintaining the local distribution network, which have nothing to do with transmission grid management. There is little incentive for an integrated grid company to pay attention to distribution grid investment. Distribution that is still closely linked to the transmission grid companies raises the risk of unfairness in power market operation; grid companies may be expected to favour their affiliates over other purchasers. A clear separation also paves the way for franchising, which allows a measure of competition, as well as the application of regulatory incentives for improved performance. Such regulatory incentives can be introduced without going so far as separating the distribution grid from the retail function.

In practice, disaggregating an integrated system raises a number of challenges, including the problem of how to reorganise distribution into a number of separate entities – and the linked issues of ownership rights and stranded assets. In China, however, the distribution networks tend to follow local government boundaries, not least because much of the investment has been driven by local governments and local generators. Separation may not be too difficult to achieve, whilst the alternative – *i.e.* keeping distribution with the transmission grid – is less helpful to meeting power sector reform objectives.

Moving toward full consumer choice

Competitive power markets need a critical mass of power purchasers. Large consumers and distribution companies are usually the first targets. In particular, distribution companies are the gateway to smaller and household consumers, pending a move to full retail competition which is likely to be some way off. They, therefore, have the capacity to make a significant contribution to emerging power markets.

Retail competition adds to the competitive forces acting on power markets, not least because it encourages the development of intermediary suppliers for end customers, who are no longer bound to their local distribution company. It requires not only the separation of distribution from transmission, but also the separation of distribution from its retail function. However, there are strong arguments for a gradual approach. If the distribution and retail functions are not completely separated (ownership unbundling), regulatory controls are needed to ensure that the distribution company does not cross-subsidise its retail activities from its network activities. There is a general need to ensure that the institutional and regulatory conditions for effectively managing wholesale competition are in place and work well, before tackling some of the issues (such as consumer protection) that will be raised by full competition[55].

If consumers do not yet have a choice of supplier – as is likely to be the case in China for the foreseeable future – a designated supplier must be identified, which is typically the distribution company. Measures need to be put in place to ensure that this supplier acts efficiently and in the best interests of the customer. Franchising is one approach.

Encouraging demand response[56]

Demand response refers to a set of strategies that aim to bring the demand side of competitive power markets into the price-setting process. It is a form of load shape management, achieved through pricing rather than command and control measures, though it also has the potential to reduce load over time. Demand-side resources are created as customers adjust their demand in response to price signals.

Demand response is viewed by some as a possible successor, at least in part, to the more highly managed and control-based demand-side management (DSM) programmes of the pre-reform era. Seen this way, programmes that require energy-efficiency actions may no longer be necessary once fully competitive markets are achieved. This is, of course, provided that consumers have the means to respond with demand-side actions, and this capability needs to be actively built into the framework for competitive markets, as early as possible.

For demand response to work, consumers must actively participate in power trading by "offering" to undertake changes in their normal behaviour patterns. Demand response relies on a range of policies that include real-time pricing, voluntary demand response programmes, direct load control, and not least, demand bidding. All consumers can participate in a demand response programme, as long as they have the flexibility to make changes to their normal electricity demand profile and to install the necessary control and monitoring technology to execute bids and demonstrate bid delivery. Consumers gain through a reduced tariff, or in some cases a financial reward via a direct payment for the electricity they did not consume at an agreed time. The concept needs a competitive power market that allows demand

55. Germany provides an example of full market opening without any specific supporting institutional and regulatory framework (no sector-specific regulator and no regulated third-party access). The experiment has been controversial – even though prices fell immediately after market opening and subsequent industry mergers consolidated *de facto* monopolies – and Germany has recently established a regulator. The EU is a different example of a process in which full consumer choice is being achieved in stages. The preamble to the 2003 Directive notes that "Electricity customers should be able to choose their supplier freely. Nonetheless a phased approach should be taken to completing the internal market for electricity to enable industry to adjust and ensure that adequate measures and systems are in place to protect the interests of customers and ensure they have a real and effective right to choose their supplier".

56. This section based on IEA (2003b).

reduction bids to compete with generation bids. Technologies must also be developed and deployed that allow customers to receive accurate price signals which form the basis of decisions to reduce consumption or shift it to off-peak periods.

In practice, high rates of demand response can be difficult to achieve. Market participants generally lack the incentive and the means to respond. Factors that typically impede demand response include: regulated retail prices, outdated metering technologies, a lack of real-time price information reaching consumers, system operators focused on supply-side resources and a historical legacy in which demand response was not considered important.

Policy intervention is therefore necessary. Demand response should be required and built directly into the structure and regulatory framework of emerging power markets. Large consumers, distribution companies, and energy service companies should all be in a position to respond quickly to high spot market prices in a way that helps all consumers save money. One policy intervention to be avoided, however, is capping prices. While price caps are well intended, they inadvertently suppress development by retailers of innovative price and service options (*e.g.* demand bids) that can enhance demand response. Capping prices can contribute to even greater problems. This was one of the strongest lessons from the California crisis and other markets that have suffered similar problems related to price level and price volatility.

If it can be achieved, demand response has a number of advantages:

- Its introduction into constrained networks will significantly dampen the price peaks often seen in wholesale markets, reducing costs and risks for all market participants. By clipping price peaks, demand response will also lead to lower wholesale prices on average and a more efficient market. This translates into real financial savings. It is estimated that incorporating demand response into the California residential market alone would lead to savings of USD 1.2 billion per year.

- It enhances energy security, especially on constrained networks, as higher concentrations of demand are typically located at network nodes, where congestion is high and network security most vulnerable.

- It helps to address the issue of market power and concentration. Market power abuses can be lowered either by reducing concentration on the supply side of the market, or by increasing the elasticity of demand relative to price. Doubling the price elasticity of demand would have the same impact on prices as halving concentration on the supply side, yet the former may be easier to achieve.

- It delivers a net reduction in consumption (*i.e.* it is not just load shifting but load reduction), which directly reduces emissions. In cases in which it simply shifts load, the environmental impact is more complex, depending on the mix of fuels and emission profiles of base load and peaking plants that have been displaced by the demand response.

Developing transparent and robust market rules

In competitive markets, delivery of efficient and effective system operation and dispatch relies on the actions of market players – in particular, the commitments they make and their delivery of these commitments. Comparing commitments to performance requires that all parties believe that accurate information on their own performance and that of others is widely available, know what the rules are, and are confident that the rules will be enforced.

Effective power markets in other parts of the world share the common feature of carefully designed market rules. A key aim of rule making is to secure a high level of transparency in market operations, as well as strong and timely information flows. This helps the market to police itself – competitors will start shouting if they notice something wrong. Specifically, rules are needed for: communication between system operators, market operators, generators, traders, etc.; scheduling and communication of bids; and information disclosure.

One effective way of setting out the rules is in a grid code. For example, the Australian National Electricity Code is a legal instrument that was developed out of a formal public consultation and approval process involving both the regulator (National Electricity Code Administrator) and the competition authority (Australian Competition and Consumer Commission).

Rules inevitably need adjustment as the market develops. The original rules may have been based on old habits and principles, and some rules can only be developed once systems are tested in real operation. To cater for this continuous adjustment, many countries have developed various fora of market participants to take issues forward. To be effective, such fora should include all important interest groups.

Choosing between different competitive power market models

As it moves towards competitive markets, China will need to take stock of the various power market models that are possible. Currently, there are more than 30 existing and planned electricity markets globally. Although each of these markets is adapted to the country or region's technical, historical and political features, they can be broadly classified according to the extent to which power market governance is centralised or decentralised. Each approach has its advantages and disadvantages. In the decentralised approach, for example, demand-side participation and demand-side price elasticity appear to be more developed. A more centralised approach, on the other hand, avoids many problems of co-ordination.

Centralised markets

The Pennsylvania-New Jersey-Maryland market is an example of a centrally controlled and operated market. Power is traded through a complex, centralised market that integrates all aspects of trading (with the possible exception of financial contracts trading) within one process. This ensures the optimal utilisation of generation resources relative to total load, taking into account transmission constraints and security requirements. A centralised system operator works out the marginal price (day-ahead and real-time market clearing prices) for each injection or withdrawal node/bus in the market, which gives rise to a large number of localised marginal prices (LMPs). These LMPs provide signals for the most efficient investments in transmission and generation.

Market participants – suppliers and consumers – can offer their resources at various price levels. They are normally free to negotiate bilateral contracts for energy, but they must work through the system operator for dispatch. The system operator organises integrated markets for trading energy and ancillary services such as day-ahead, hour-ahead and balancing/real-time markets. The system operator is also the market operator.

This market model is well suited to a power sector in which the system operator has, or is given, tight control of all physical resources in the market, and can then effectively optimise their utilisation. However, the model has been criticised for its focus on generation resources, and for raising problems in relation to demand response. Market participants have significant constraints on their freedom. They cannot, for example, control the dispatch of their own resources.

Decentralised markets

In decentralised markets – NordPool is an example – market participants are given far more control of their own resources. The market operator and system operator are separate. In principle, the market operator is responsible only for facilitating the trade of energy as the commodity, but within the physical constraints set by the system operator. The operation of the physical system is the sole responsibility of the system operator.

In addition, the market participants are given the freedom and responsibility of controlling (scheduling) their resources, and to self-manage optimized utilisation of their physical and contractual assets. An important, and necessary, principle is that of balance responsibility, meaning that the market participants must obtain a net portfolio balance in advance of real-time operation. Any deviation from net balance must be compensated for in the real-time balancing market.

Other types of markets

Other market models typically fall between these two extremes. For example, California's power exchange is an extended PJM-style structure with an additional intra-day market in addition to the day-ahead and real-time energy market. Single price, centrally dispatched markets are simplified versions of the PJM model without the day-ahead market (examples are the first deregulated markets in the world, such as Chile, Argentina, Brazil, Australia and New Zealand, as well as the first United Kingdom pool). Some of these markets allow physical bilateral contracts market, while others – such as Australia's NEMMCO – allow only financial hedges. At the other extreme, with respect to bilateral contracts, is the new United Kingdom market (NETA), in which the bulk of power is traded on bilateral contracts and though more or less formal exchanges, and where the only government controlled market is a balance market.

The European exchanges tend towards the decentralised end of the spectrum. All are based on a separation between the market operator and the system operator, although in most cases the market operator is owned by the system operator. The main differences are whether these markets have implemented the principle of balanced schedules, and whether the bids and offers to the power exchange (the day-ahead market) are portfolio-based or unit-based. The Spanish and Italian markets use unit-based bidding, which means that generation offers are connected to physical generation units. In this structure, the market operator performs a type of centralised dispatch – or, at least, adjusts bids and offers in the day-ahead market in order to relieve transmission congestion. Most other European markets, including PowerNext and NordPool among others, are portfolio-based and leave scheduling to the market participants. Congestion is handled either by market splitting, or more commonly by auction of transmission rights across borders. The system operator handles congestion in real time through special regulation such as counter trade.

CONCLUDING REMARKS

China's power sector reform achievements to date are considerable. Building on these achievements through further, carefully managed reform can be expected to unlock the potential that exists for a considerably more efficient and less polluting power sector. It is important to sustain the reform momentum, and to consolidate the changes that were started some 20 years ago. The power sector presents many complex features, and the holistic approach to reform advocated in this report – ensuring that energy efficiency and environmental goals as well as economic efficiency are promoted together – is critical if China is to secure an effective and sustainable long-term outcome. Managing demand while strengthening supply holds the best prospects for mitigating supply/demand imbalances. China's continuing dependence on coal as the main fuel input to power generation requires a robust approach to tackling pollution. It is vital to ensure that reform helps, rather than hinders, this.

A holistic approach also implies the need for all aspects of the reform process to be developed and consolidated. Effective power sector reforms rule out the option of "picking and choosing", by leaving out or delaying for too long specific reforms that are essential to the success of the whole process. One can think of a complex regulatory system as being much like a complex mechanical system, such as an automobile; fuel, engine, and wheels are all essential, but without a transmission the wheels won't go round, and without a steering wheel the vehicle won't get far. Building competitive markets and making prices cost-reflective are key elements of a reform effort, but without well-designed regulations, without institutions for implementing and monitoring, and without a strategic plan to guide them, there is no assurance it will arrive at the desired endpoint.

China's important structural reforms, which have already established a diverse and potentially strong set of market players for the development of competition, now need backing up with the further reforms discussed in this report. Sequencing is important, and this report strongly advocates that efforts should, in the first instance, be directed toward institutional and pricing reforms. The current efforts to test competition in some regions cannot progress very far, if at all, without a strong regulatory framework and a more soundly based, cost-reflective pricing policy. This report also encourages China to consider whether it can take some basic steps towards cross-regional or provincial trading on a competitive basis.

All these actions could begin to realise near-term efficiency benefits, by getting more out of existing infrastructure, as well as by improving understanding of a new market-based framework and moving China away from the old planning traditions. An overarching message of this report is the importance of transparency, and the pressing need for China to find ways of improving transparency – institutionally at all levels of government, in the power sector's corporate structures, and across the emerging regulatory framework as a whole.

China does not need to travel alone on the road to reform. Although the country must take its own unique path to fit its particular situation, the reform experiences – both positive and negative – of other countries are very instructive. These experiences can shed light not only on the elements of reform, but also on the process itself, and the various ways in which reform can be taken forward and promoted.

The IEA looks forward to working in co-ordination with the other organisations already involved in power sector issues in China, with the aim of drawing upon these experiences so that China can identify the most effective approaches, which suit its circumstances. The IEA also looks forward to learning from China's experience, and sharing lessons it holds for power sector reforms in other countries. Across the world, power sector reform is a continuing process; no country can claim to have worked it all out.

Drawing on relevant experiences elsewhere, it is possible to identify a number of specific areas that might be useful for developing a more detailed dialogue with China. This report already covers most of these at a broad level; in most cases, the need now is to be more specific and to develop the proposed design. The fields in Box 26 are offered as suggestions, in no particular order, of topics that the IEA would be willing to explore with China. In the end, of course, it is up to China to determine how it can most effectively deploy its resources as it determines the next steps to be taken in continuing to reform its power sector.

Box 26 Potential fields of future international collaboration in power sector reform

- Legal groundwork and the revisions to the electricity law, to ensure that the best and most relevant of experiences elsewhere can be reviewed and tested against China's needs.

- Data collection and analysis, to secure a stronger understanding of power sector supply/demand developments.

- Institutional development, to reinforce SERC's capacities as regulator. This report already makes a few suggestions: linking up with regulators' clubs, considering the experience of particular regulators whose development may share common points with China, and improving understanding and competences for tackling anti-competitive behaviour, again by linking up with relevant experts elsewhere.

- Environmental regulation, to bolster incentives designed to encourage developers to employ more efficient, less-polluting coal technologies, as well as cleaner fuels, and to increase the transparency of environmental costs to electricity consumers.

- Pricing, to develop appropriate methods for cost-based pricing that distinguishes between various services. The design and implementation of lifeline support for vulnerable consumers also need development.

- Investment approval regimes, to strengthen grid planning and investment to reflect costs more clearly, and to ensure that environmental considerations are taken into account.

- Investment environment, to gauge the potential for encouraging more private sector involvement and foreign investment.

- Demand-side management, to review and enhance current practices, *e.g.* financial and other incentives for investing in energy efficiency. China may wish to consider taking part in the IEA's Implementing Agreement on DSM.

- Further structural reforms to ensure the independence of the grid and system operation, and to improve corporate governance across the sector, including generating companies.

- System security, to draw lessons from best practices elsewhere.

- Development of some cross-regional/provincial competition, to gain experience with competitive power markets. This would require a particularly careful review of China's specific circumstances and of the experiences elsewhere that fit it best, as well as identifying all the actions that would be necessary for it to work (such as dealing with current PPAs).

- Rural electrification, to improve welfare nationwide. The task of bringing power to all of China's citizens is not covered at all in this report, but is clearly of great importance. It links to issues such as designing support for the poorest members of society and grid planning, among others.

ABBREVIATIONS

CCT	clean coal technology
CEER	Council of European Energy Regulators
CfD	contracts for differences
CSG	China Southern Power Grid Company Limited
DSM	demand side management
ESCO	energy service company
ETSO	European Transmission System Operators
FGD	flue gas desulphurisation
HHI	Herfindahl-Hirschman index
IA	Implementing Agreement
IEA	International Energy Agency
IGCC	integrated gasification and combined cycle generation
IPP	independent power producer
IRP	integrated resource planning
ISO	independent system operator
LMP	localised marginal price
LRIC	long run incremental cost
NDRC	National Development and Reform Commission
NEM	National Electricity Market (Australia)
NEMMCO	National Electricity Market Management Company (Australia)
OECD	Organisation for Economic Co-operation and Development
PJM	Pennsylvania-New Jersey-Maryland (power market)
PPA	power purchase agreement
RMB	*Renminbi* (name of China's currency)
SASAC	State-owned Assets Supervision and Administration Commission
SEPA	State Environmental Protection Administration
SERC	State Electricity Regulatory Commission
SETC	State Economic and Trade Commission
SG	State Grid Corporation of China
SME	small and medium enterprises
SOE	state-owned enterprise
SPC	State Planning Commission
TOU	time of use (pricing)
TSO	transmission system operator
T&D	transmission and distribution
WTO	World Trade Organisation

BIBLIOGRAPHY

Andrews-Speed, P. (2003), *Energy Policy and Regulation in the People's Republic of China*, Kluwer Law International, The Hague.

Berrah, N., R. Lamech and J.P. Zhao, (eds.) (2001), *Fostering Competition in China's Power Markets*, World Bank Discussion Paper 416, The World Bank, Washington, DC.

Berrah, N. and J. Wright, (2002), *Establishment of a State Electricity Regulatory Commission in China: A Suggested "Roadmap"*, World Bank Working Paper 33209, The World Bank, Washington, DC.

China Energy Development Strategy and Policy Research Group (2004) *Research on National Energy Comprehensive Strategy and Policy of China*, Economic Science Press, Beijing.

Editorial Board of the China Electricity Yearbook (2004), *China Electricity Yearbook 2004*, China Electricity Press, Beijing.

Editorial Board of the China Electricity Yearbook (2004), *China Electricity Yearbook*, China Electricity Press, Beijing.

Energy Conservation Dissemination Centre (2006), *National Electric Power Industry Statistical Report*, Beijing (available at www.secede.org.cn/newscontent.asp?id=360).

Energy Foundation, China Sustainable Energy Programme (2002), *Strategies for China's Electricity Reform and Renewable Development*, The Energy Foundation, San Francisco.

Energy Foundation, China Sustainable Energy Programme (2003), *Demand-side Management in China*, The Energy Foundation, San Francisco.

General Office of the China Secretary of the State Council (2002), *The Electrical Power Reform Scheme*, General Office of the China Secretary of the State Council, Beijing.

Green, R.J. (1997), "Electricity Transmission Pricing: An International Comparison", *Utilities Policy*, Vol. 6, No 3, pp. 177-184.

Hu, Z. *et al.* (2005), *Demand-side Management in China's Restructured Power Industry – How Regulation and Policy Can Deliver Demand-side Management Benefits to a Growing Economy and a Changing Power System*, Energy Sector Management Assistance Project (ESMAP) and World Bank, Washington, DC.

Hunt, S. (2002), *Making Competition Work in Electricity Markets*, John Wiley, New York.

Hunt, S. and G. Shuttleworth (1996), *Competition and Choice in Electricity*, John Wiley, Chichester.

International Energy Agency (IEA) (2001), *Regulatory Institutions in Liberalised Energy Markets*, IEA/Organisation for Co-operation and Development (OECD), Paris.

IEA (2002), *Security of Supply in Electricity Markets*, IEA/OECD, Paris.

IEA (2003a), *Power Generation Investment in Electricity Markets*, IEA/OECD, Paris.

IEA (2003b), *The Power to Choose – Demand Response in Liberalised Electricity Markets*, IEA/OECD, Paris.

IEA (2003), *World Energy Investment Outlook*, IEA/OECD, Paris.

IEA (2004), *World Energy Outlook*, IEA/OECD, Paris.

IEA (2005a), *Lessons from Liberalised Electricity Markets*, IEA/OECD, Paris.

IEA (2005b), *Learning from the Blackouts – Transmission System Security in Competitive Electricity Markets*, IEA/OECD, Paris.

IEA (2005c), *Saving Electricity in a Hurry*, IEA/OECD, Paris.

National Bureau of Statistics (NBS) (2001), *China Energy Statistical Yearbook, 1997-1999*, China Statistics Press, Beijing.

NBS (2004), *China Energy Statistical Yearbook, 2000-2002*, China Statistics Press, Beijing.

NBS (2005a), *China Energy Statistical Yearbook, 2004*, China Statistics Press, Beijing.

NBS (2005b) *China Statistical Yearbook 2005*, China Statistics Press, Beijing.

NBS (2006) *Economic and Social Statistical Report of the People's Republic of China 2005*, NBS, Beijing, (available at www.stats.gov.cn).

OECD (2003), *OECD Investment Policy Reviews – China, Progress and Reform*, OECD, Paris.

OECD (2004), *Principles on Corporate Governance*, OECD, Paris.

OECD (2005a), *China in the Global Economy – Governance in China*, OECD, Paris.

OECD (2005b), *OECD Economic Survey – China*, OECD, Paris.

Philibert, C. and J. Podkanski (2005), *International Energy Technology Collaboration and Climate Change Mitigation – Case Study 4: Clean Coal Technologies*, COM/ENV/EPOC/IEA/SLT (2005) (4), IEA/OECD, Paris.

The Regulatory Assistance Project (2000), *Best Practices Guide: Implementing Power Sector Reform*, The Regulatory Assistance Project, Gardiner, Maine.

The Regulatory Assistance Project (2002), *Options for the Institutional Reform of China's Electrical Power Industry*, The Regulatory Assistance Project, Gardiner, Maine.

The Regulatory Assistance Project (2003), *Report on the Establishment of the State Electricity Regulatory Commission*, The Regulatory Assistance Project, Gardiner, Maine.

Shao, S.W., Z.Y. Lu, N. Berrah, B. Tenenbaum, and J.P. Zhao (eds.), (1997), *China: Power Sector Regulation in a Socialist Market Economy*, World Bank Discussion Paper 361, The World Bank, Washington, DC.

Statskontoret (Swedish Agency for Public Management), 2005, *Competition at the Public/Private Interface*, Statskontoret, Stockholm.

Vandenbergh, F. (2004), "Development of Interconnections and Reliability Standards", in J. Bielecki and M.G. Desta (eds.), *Electricity Trade in Europe. Review of the Economic and Regulatory Challenges*, Kluwer Law, The Hague.

World Bank (1994), *China – Power Sector Reform: Towards Competition and Improved Performance*, World Bank Report 12929-CHA, The World Bank, Washington, DC.

World Bank (2000), *The Private Sector and Power Generation in China*, World Bank Discussion Paper 406, The World Bank, Washington, DC.

World Bank (2004), *Reforming Infrastructure: Privatisation, Regulation and Competition*, World Bank Policy Research Report, The World Bank, Washington, DC.

World Bank and Energy Foundation (2000), *Workshop: New Waves of Power Sector Reform in China*, proceedings of workshop in Beijing, The World Bank, Washington, DC.

Xu, Y. (2002), *Powering China. Reforming the Electric Power Industry in China*, Ashgate, Burlington.

Ye, R.S. *et al.* (eds.) (2004 and 2005), *Electric Power in China*, China Electric Power Information Centre, Beijing.